A REFERENCE GUIDE TO PRACTICAL ELECTRONICS

A REFERENCE GUIDE TO PRACTICAL ELECTRONICS

Robert G. Krieger, Sr.

Wayne Community College
Goldsboro, North Carolina

McGraw-Hill Book Company
Gregg Division

New York Atlanta Dallas St. Louis San Francisco Auckland Bogotá
Guatemala Hamburg Johannesburg Lisbon London Madrid Mexico
Montreal New Delhi Panama Paris San Juan São Paulo Singapore
Sydney Tokyo Toronto

Library of Congress Cataloging in Publication Data

Krieger, Robert G
  A reference guide to practical electronics.

  Bibliography: p.
  1. Electronics. I. Title.
TK7816.K74   621.381   80-14812
ISBN: 0-07-035492-8

A Reference Guide to Practical Electronics

  2 3 4 5 6 7 8 9 0 MUMU 8 8 7 6 5 4 3 2

Sponsoring Editor: Mark Haas
Editing Supervisor: Katharine Glynn
Design Supervisor: Nancy Axelrod
Production Supervisor: Priscilla Taguer

Cover Illustration: Jon Weimar

CONTENTS

AC Circuits

Communications

PREFACE

A Reference Guide to Practical Electronics offers a highly organized and condensed overview of electronics. One hundred carefully selected equations function as keys to five major areas via explanations that are straightforward and to the point. Material covered in this fashion provides the electronics student with an effective supplementary learning tool.

On occasion, the student will encounter an impassable problem while doing assignments out of class; in many cases, reference to this text should clarify the issue and allow progress to resume. It should also be noted that the information within each section is not simply scattered but progresses logically from page to page. This volume should be retained in the graduate's library as a future source of highly concentrated data for on-the-job applications, review for the FCC exam, and personal hobby uses.

Robert G. Krieger, Sr.

DC CIRCUITS

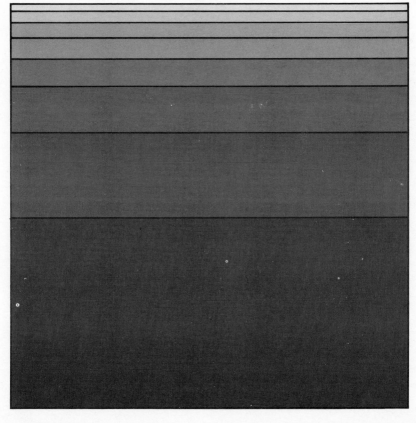

$$I = \frac{Q}{T}$$

I = current   A
Q = charge   C
T = time   s

The intensity of current flow in any circuit results from the volume of electrons that flow past a point during a given time period. To understand this process completely, you have to understand first that the unit of charge, designated Q, is measured in coulombs (C), one coulomb equaling 6.25×10^{18} electrons. Thus, if you had 6.25×10^{18} electrons of charge deposited upon a metal plate, there would be a standing charge of 1 C. By examining the equation, you can see that current is not simply a volume of electrons but a flow of electrons per unit of time. If 1 C of charge were to revolve around and pass the same point in a circuit every second(s), a current of one ampere (1 A) would result. If 20 C passed that same point every second, a current of 20 A would result. Increasing the charge produces a greater current, while increasing the time factor will decrease the current as follows:

$$I = \frac{Q}{T}$$

Therefore
$$\frac{50\text{ C}}{5\text{ s}} = 10\text{ A}$$

while
$$\frac{50\text{ C}}{10\text{ s}} = 5\text{ A}$$

and
$$\frac{100\text{ C}}{5\text{ s}} = 20\text{ A}$$

It is of interest to note that the letter I stands for *intensity*, while the units are in amperes (A). The direction of current flow is from the negative terminal of the power source, through the circuit, and toward the positive terminal. This direction is referred to as *electron flow*.

EXAMPLE 1-1

Refer to Fig. 1-1 and find the current if 30 C passes point A every second.

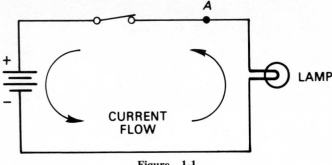

Figure 1-1

$$I = \frac{Q}{T}$$

$$= \frac{30\text{ C}}{1\text{ s}}$$

$$= 30\text{ A}$$

EXAMPLE 1-2

Refer to Fig. 1-2 and find the current if 3 C passes point A every 20 s.

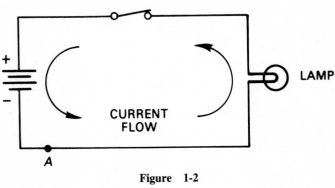

Figure 1-2

$$I = \frac{Q}{T}$$

$$= \frac{3\text{ C}}{20\text{ s}}$$

$$= 0.15\text{ A} = 150\text{ mA}$$

Note It does not matter at what point in the circuit the measurement is taken, since current flow in a closed loop is the same throughout.

EQUATION 2
CHARGE

$$Q = I \times T$$

Q = charge	C
I = current	A
T = time	s

From the basic formula for current (Equation 1) we can derive the statement that the accumulated charge at a point is the current times the length of time that it is sustained. Note that in a closed loop the electrons are in constant motion and cannot build up. This is *current*. For a charge to accumulate, the circuit must be broken at some point, as through a capacitor. The capacitor consists of two metal plates separated by a dielectric, such as air, plastic, or another insulating material, in which the charge Q accumulates.

The equation indicates that if either the time or current factor is increased, the overall charge will also increase, as follows:

$$Q = I \times T$$

Therefore	$3\,A \times 4\,s = 12\,C$
while	$6\,A \times 4\,s = 24\,C$
and	$3\,A \times 10\,s = 30\,C$

EXAMPLE 2-1

Refer to Fig. 2-1 and determine what charge will accumulate in a capacitor if a current of 2 A is sustained for 10 s.

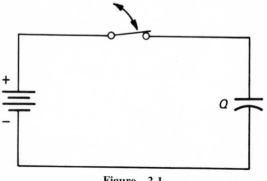

Figure 2-1

$$Q = I \times T$$
$$= 2\,\text{A} \times 10\,\text{s}$$
$$= 20\,\text{C}$$

EXAMPLE 2-2

What charge will accumulate in the capacitor if a current of 35 microamperes (μA) is sustained for 360 milliseconds (ms)?

$$Q = I \times T$$
$$= (35 \times 10^{-6})\,\text{A} \times (360 \times 10^{-3})\,\text{s}$$
$$= 1.26 \times 10^{-5}\,\text{C}$$

Note In Example 2-2, the resultant answer may not seem to represent a large number of electrons. If we multiply this result by the number of electrons in one coulomb, however, we can see that the small charge of 1.26×10^{-5} C equals 7.9×10^{13} electrons.

EQUATION 3
CURRENT (OHM'S LAW)

$$I = \frac{V}{R}$$

I = current   A
V = voltage   V
R = resistance   Ω

Perhaps the most fundamental and useful equation is Ohm's law. It states the relationship between the three main circuit elements: current, voltage, and resistance. It states, you will notice, that if either the voltage or the resistance in a circuit is changed, the current must also change. More specifically, if the applied voltage in a circuit with a lamp is held constant and the lamp is replaced with another of lower resistance, the current will increase. If the original lamp is left in the circuit and the applied voltage is increased, an increase in current will also result. Lowering the voltage or increasing the resistance has just the opposite effect on the current.

When working within an actual current you will often find it more convenient to take a voltage reading across a resistor. You can calculate the current from that, instead of having to break the resistor connection to measure current. As in most basic equations, the values entered must be in base units in order that the equation generate the base unit, the ampere, as an answer. Of course, multiplying the voltage and resistance terms by the same factor will also serve this same purpose. For example, the division of kilovolts by kilohms yields the same answer in base units of amperes.

EXAMPLE 3-1

Refer to Fig. 3-1 and find the circuit current when a 6-ohm (6-Ω) lamp* is connected to a 12-volt (12-V) car battery.

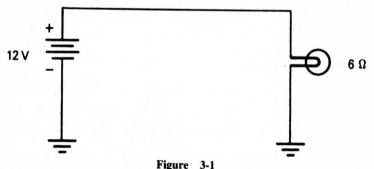

Figure 3-1

*In this and other examples in this text, tungsten lamp resistances are given as fixed values for simplicity. In reality, they increase somewhat with increased potential applied.

$$I = \frac{V}{R}$$

$$= \frac{12\ \text{V}}{6\ \Omega}$$

$$= 2\ \text{A}$$

EXAMPLE 3-2

What will the current be if, instead of a single 12-V battery, two are used, as shown in Fig. 3-2?

Figure 3-2

$$I = \frac{V}{R}$$

$$= \frac{24\ \text{V}}{6\ \Omega}$$

$$= 4\ \text{A}$$

Note Although the lamp in Example 3-2 will be much brighter as a result of the greater current flow, the batteries will have only one-half the life, since they must sustain the heavier current.

EQUATION 4 VOLTAGE			
$V = I \times R$	V = voltage	V	
	I = current	A	
	R = resistance	Ω	

This variation of Ohm's law states that if there is a current through a resistance, a voltage must be present across it to have produced the current. Furthermore, the voltage present is the direct product of the current and the resistance itself. If either is increased, the potential difference across the resistance must be greater.

Using this equation, we can now make two interesting points. First, if a power source directly drives a single resistive load and that resistance is increased, the voltage across the load remains the same and only the current changes. On the other hand, if, as in Fig. 4-1, there is a second resistance

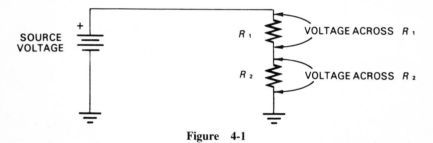

Figure 4-1

and the ohmic value of R_2 is increased, the voltage across R_2 will also increase. This extra voltage that is "picked up" is transferred from R_1, and thus R_1 now has less voltage across it. If R_2 is decreased to 0 (a dead short circuit), there cannot be any voltage across it, and the entire source voltage is then across R_1. It is the second resistor in this circuit that changes the voltage across the first; in contrast, in the first example the entire source voltage was applied to the single load. Studying Equation 5 for resistors in series will aid you in understanding the above examples.

EXAMPLE 4-1

Determine the voltages across each of the resistors, R_1 and R_2, shown in Fig. 4-2.

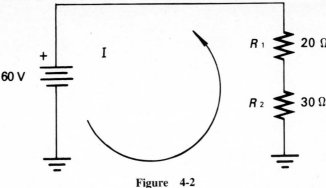

Figure   4-2

Before beginning your analysis and calculations, note that R_2 is the greater of the two resistances and should have the higher potential across it.

From Equation 3,

$$I = \frac{V}{R} = \frac{60 \text{ V}}{50 \text{ } \Omega} = 1.2 \text{ A} \qquad \textbf{total circuit current}$$

Therefore

$$V_{R_1} = I \times R_1 = 1.2 \text{ A} \times 20 \text{ } \Omega = 24 \text{ V}$$

$$V_{R_2} = I \times R_2 = 1.2 \text{ A} \times 30 \text{ } \Omega = 36 \text{ V}$$

As a double check, these two voltage drops added together should equal the source voltage:

$$\textbf{24 V + 36 V = 60 V}$$

EQUATION 5
RESISTANCE, SERIES

$$R_T = R_1 + R_2 + \cdots + R_n$$

R_T = total resistance $\quad\Omega$

$R_1 \ldots R_n$ = each resistance $\quad\Omega$

The total resistance of any given series circuit equals all circuit resistances added together. Almost every electronic component contains a certain amount of resistance. The term *resistance* refers to the opposition to current flow. For example, a 50,000-Ω resistor will resist current flow to a much greater degree than will a 10-Ω unit.

One way to demonstrate this is to add resistance into a functioning battery-lamp circuit. The result is a dimmer lamp. The additional resistance restricts the flow of current and (from Equation 4) reduces the voltage across the lamp. Adding more resistors in series with the lamp produces a greater total resistance. This, in turn, reduces the current eventually to a point where the lamp appears to go out.

Examples of resistive elements are lamps, resistors themselves, wire, and even the internal portion of batteries, to mention only a few. Resistors are manufactured in a wide range of resistances and power ratings (wattages), so that we can create any combination of needed circuit voltages and currents. Short lengths of heavy-gage wire may be considered to have zero resistance for most applications.

EXAMPLE 5-1

Determine the voltage across the lamp shown in Fig. 5-1.

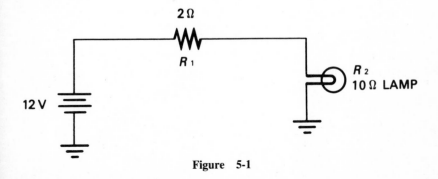

Figure 5-1

First, find the total circuit resistance.

$$R_T = R_1 + R_2$$
$$= 2\,\Omega + 10\,\Omega = 12\,\Omega$$

Next, find the total current.

$$I_T = \frac{V}{R} = \frac{12\,V}{12\,\Omega} = 1\,A$$

Finally, $\qquad V_{LAMP} = I \times R = 1\,A \times 10\,\Omega = 10\,V$

EXAMPLE 5-2

Now, if you add another 2-Ω resistor R_3 to the circuit as in Fig. 5-2, you notice two things: the total resistance increases, and the lamp voltage decreases.

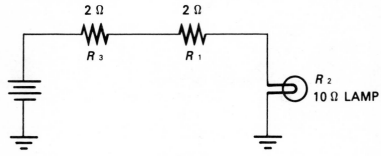

Figure 5-2

$$R_T = R_1 + R_2 + R_3$$
$$= 2\,\Omega + 10\,\Omega + 2\,\Omega = 14\,\Omega$$

Thus, $\qquad I_T = \frac{V}{R} = \frac{12\,V}{14\,\Omega} = 0.857\,A$

and $\qquad V_{LAMP} = I \times R = 0.857\,A \times 10\,\Omega = 8.57\,V$

EQUATION 6
RESISTANCES, TWO IN PARALLEL

$$R_T = \dfrac{R_1 \times R_2}{R_1 + R_2}$$	R_T = total parallel resistance	Ω
	R_1 = first resistance	Ω
	R_2 = second resistance	Ω

This particular form of the equation for parallel resistance is applicable when exactly two resistances are placed in parallel with each other and it is desired to find their combined (equivalent) resistance. Before proceeding with its use, always consider this fact: whatever the answer obtained, it must be less in value than the lower of the two resistances. For example, when you parallel a 5-Ω resistor with a 50-Ω unit, the result must be less than 5 Ω. This provides us with somewhat of a check. If the answer is higher than the lower resistance, an error has been made.

Obtaining the combined resistance value of resistances in parallel also greatly simplifies other circuit calculations, as can be seen in Fig. 6-1.

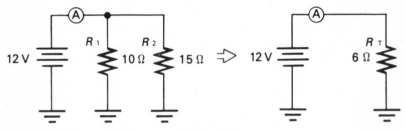

Figure 6-1

$$R_T = \frac{R_1 \times R_2}{R_1 + R_2}$$

$$= \frac{10 \times 15}{10 + 15}$$

$$= 6 \, \Omega$$

$$I = \frac{V}{R} = \frac{12 \, V}{6 \, \Omega} = 2 \, A$$

In finding the total circuit current, you can see that it splits through R_1 and R_2. By combining the two resistances into one with Equation 6, you can use the basic formula $I = V/R$ and obtain the solution of 2 A.

Lamps and other resistive loads also work directly into this formula, as do actual resistors themselves. For example, paralleling a 50-Ω kitchen toaster with a 200-Ω lamp would produce a 40-Ω total load.

EXAMPLE 6-1

Find the total circuit resistance of the circuit shown in Fig. 6-2 with the aid of Equation 6.

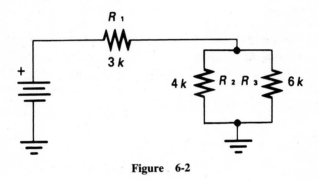

Figure 6-2

The result of the parallel combination is that

$$R_{2,3} = \frac{(4 \times 10^3) \times (6 \times 10^3)}{(4 \times 10^3) + (6 \times 10^3)} = \frac{24 \times 10^6}{10 \times 10^3} = 2400 \ \Omega$$

The resulting series combination can then be found.

$$R_T = R_1 + R_{2,3} = 3 \ k\Omega + 2400 \ \Omega = 5400 \ \Omega$$

EXAMPLE 6-2

Find the total circuit resistance of the circuit shown in Fig. 6-3 with the aid of Equation 6.

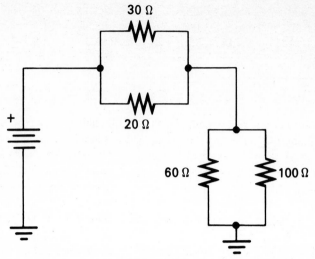

Figure 6-3

Note Find the two parallel combinations separately and then add them together.

$$R_{T_1} = \frac{30 \times 20}{30 + 20} = \frac{600}{50} = 12\ \Omega$$

$$R_{T_2} = \frac{60 \times 100}{60 + 100} = \frac{6000}{160} = 37.5\ \Omega$$

$$R_T = R_{T_1} + R_{T_2} = 12\ \Omega + 37.5\ \Omega$$

$$= 49.5\ \Omega$$

EQUATION 7
RESISTANCES, MORE THAN TWO IN PARALLEL

$$R_T = \frac{1}{1/R_1 + 1/R_2 + \cdots + 1/R_n}$$

R_T = total resistance $\quad \Omega$

$R_1 \ldots R_n$ = each resistance $\quad \Omega$

When only two resistances are in parallel, Equation 6 may be used. If three or more resistances are in parallel, Equation 7 must be used. An examination of the equation reveals that as more and more resistances are added in parallel, the overall circuit resistance becomes lower. Also, this total resistance must be less than the value of any of the parallel resistances. For example, paralleling a 5-Ω, a 10-Ω, and a 15-Ω resistor results in a total resistance of 2.7 Ω, which is considerably less than the lowest single resistance value.

As with any calculation using reciprocals (i.e., numbers in the form $1/x$), care must be taken not to round off decimals to excess before they are divided into the number 1, since this will result in inaccurate answers. Rounding off to the second decimal place is recommended and, with any simple calculator, poses no problem.

Term R_n in the equation means that any number of additional resistances may be added into the calculation. For instance, if five resistors are in parallel, then the equation is.

$$R_T = \frac{1}{1/R_1 + 1/R_2 + 1/R_3 + 1/R_4 + 1/R_5}$$

EXAMPLE 7-1

In Fig. 7-1, a 12-V car battery drives three resistive loads in parallel: a car radio of 30 Ω, a trouble lamp of 20 Ω, and a small electric heater of 6 Ω. What is the total load resistance that the battery must drive?

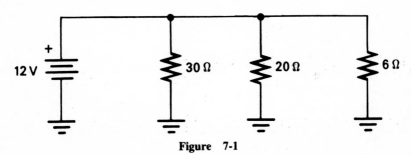

Figure 7-1

$$R_T = \frac{1}{1/R_1 + 1/R_2 + 1/R_3}$$

$$= \frac{1}{1/30 + 1/20 + 1/6}$$

$$= \frac{1}{0.03 + 0.05 + 0.17}$$

$$= \frac{1}{0.25}$$

$$= 4\,\Omega$$

EXAMPLE 7-2

In Fig. 7-2, what is the total circuit current flow?

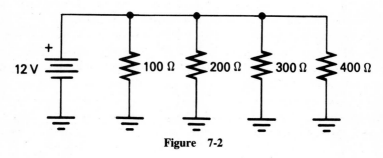

Figure 7-2

First, find the total load resistance of the four resistors connected in parallel.

$$R_T = \frac{1}{1/R_1 + 1/R_2 + 1/R_3 + 1/R_4}$$

$$= \frac{1}{1/100 + 1/200 + 1/300 + 1/400}$$

$$= \frac{1}{0.02083}$$

$$= 48\,\Omega$$

Then, using Equation 3, find the current.

$$I = \frac{V}{R} = \frac{12}{48} = 0.25\text{ A}$$

Note The total resistance of 48 Ω is less than the lowest-value single resistor, which is 100 Ω.

EQUATION 8
POWER: VOLTAGE AND CURRENT

$P = V \times I$	P = power	W
	V = voltage	V
	I = current	A

An electric device consumes 1 watt (W) of energy when an applied voltage of 1 V produces a current flow of 1 A. Power consumption is usually evidenced by the giving off (dissipation) of heat. The greater the wattage consumed, the greater the heat dissipated. In a circuit, if either the voltage or the current is increased, an increase in power will result. In the first case, an increase in potential (voltage) can be provided by simply increasing the number of voltaic cells. In the second, lowering the load resistance will result in a current flow increase. This second possibility exists, however, only if the driving source (battery) has a very low internal resistance and is a strong source of current. Should the driving source have more than a very little internal resistance, then for maximum power to be developed, the load itself should equal the ohmic value of the internal resistance.

This basic DC power equation is equally applicable to an AC circuit in determining the operating cost of power for a given period of time. The equation is first applied to determine the actual amount of wattage being consumed. This wattage is then converted to kilowatts and multiplied by the length of usage to give kilowatthours (kWh). Finally, this kWh total is multiplied by the electric rate (cost per kWh) for your area. This type of calculation can be seen in Example 8-2.

EXAMPLE 8-1

Show that the circuit in Fig. 8-1*b* consumes more power than the circuit in Fig. 8-1*a*.

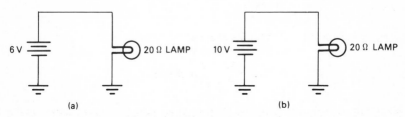

(a)   (b)

Figure 8-1

For Fig. 8-1a

$$I = \frac{V}{R} = \frac{6}{20} = 0.3\,\text{A}$$

$$P = V \times I = 6 \times 0.3 = 1.8\,\text{W}$$

For Fig. 8-1b

$$I = \frac{V}{R} = \frac{10}{20} = 0.5\,\text{A}$$

$$P = V \times I = 10 \times 0.5 = 5\,\text{W}$$

Thus, the circuit in Fig. 8-1b consumes 3.2 W more than the circuit in Fig. 8-1a.

EXAMPLE 8-2

What is the cost of operating a lamp that draws 9 A at 120 V AC for 16 hours (h) if the electric rate is \$0.05/kWh?

$$P = V \times I = 120 \times 9 = 1080\,\text{W} = 1.08\,\text{kW}$$

$$1.08\,\text{kW} \times 16\,\text{h} = 17.28\,\text{kWh}$$

$$17.28\,\text{kWh} \times \$0.05 = \$0.86$$

EQUATION 9 POWER: CURRENT AND RESISTANCE		
$P = I^2R$	P = power	W
	I = current	A
	R = resistance	Ω

The second of the three power equations is useful when only the circuit current and resistance are given. It is obvious that the voltage can be calculated and then Equation 8 can be applied. This, however, is not the fastest way, and in every additional calculation there enters another chance for error.

Equation 9 also tells us that power is a direct function of the square of the current. In other words, if the current in a circuit is doubled, the power will increase by a factor of four or 4 times.

$$P = I^2 \times R$$

If I = 3 A and R = 10 Ω:

$$P = 3^2 \times 10$$
$$= 9 \times 10$$
$$= 90 \text{ W}$$

If I = 6 A and R = 10 Ω:

$$P = 6^2 \times 10$$
$$= 36 \times 10$$
$$= 360 \text{ W}$$

Note that doubling the current of 3 A to 6 A causes the power to increase 4 times, from 90 to 360 W.

EXAMPLE 9-1

A current meter indicates that 40 mA of current is flowing in the circuit of Fig. 9-1. What should the wattage rating of the resistor be so that the resistor will not overheat and burn out?

Figure 9-1

$$P = I^2R$$
$$= 0.04^2 \times 500$$
$$= 0.0016 \times 500$$
$$= 0.8 \text{ W}$$

The calculated power consumption is 0.8 W. A general rule of thumb suggests that this value be doubled as a safety factor.

$$2 \times 0.8 \text{ W} = 1.6 \text{ W}$$

The closest commercial value to this would be a 2-W resistor.

EXAMPLE 9-2

A portable electric heater bears a tag that states that the unit draws 9 A of current. An ohmmeter reading* taken at the plug indicates that the resistance of the element is 13 Ω. What is the power consumption of this unit?

$$P = I^2R$$
$$= 9^2 \times 13$$
$$= 81 \times 13$$
$$= 1053 \text{ W}$$

*Ohmmeter readings should be taken on devices only when they are unplugged or removed from any source of power.

EQUATION 10
POWER: VOLTAGE AND RESISTANCE

$$P = \frac{V^2}{R}$$

P = power	W
V = voltage	V
R = resistance	Ω

Equation 10, the third and final version of the power equation, states that power may be expressed as the voltage squared divided by the resistance. This particular version is quite useful, since it does not require the current value to calculate power. This advantage becomes clearer when it is realized that we do not have to break the circuit for a particular current reading. We need only a direct voltage and ohmage reading, and the calculations are performed from them. Breaking an electric circuit is quite difficult if there is a confining parts layout; also, the parts themselves may be distorted or damaged in the cutting process. Not having to have the current value is also advantageous in AC circuits, since most basic volt-ohm-milliammeters (VOMs) are not capable of measuring AC current but do measure AC voltage and resistance. From these last two values the AC power may be determined.

EXAMPLE 10-1

What power is being dissipated by resistor R_3 in Fig. 10-1?

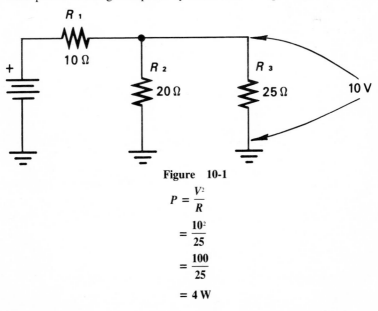

Figure 10-1

$$P = \frac{V^2}{R}$$

$$= \frac{10^2}{25}$$

$$= \frac{100}{25}$$

$$= 4 \text{ W}$$

EXAMPLE 10-2

In Fig. 10-2, what is the wattage of each lamp? Note: Because lamps have equal resistances, the voltage across each lamp is one-half the line voltage.

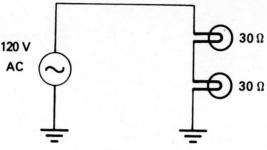

Figure   10-2

$$P = \frac{V^2}{R}$$

$$= \frac{60^2}{30}$$

$$= \frac{3600}{30}$$

$$= 120 \text{ W} \quad \text{each lamp}$$

EQUATION 11
VOLTAGE DIVIDER

$$V_{OUT} = \frac{R_X}{R_T} \times V_S$$

V_{OUT} = output voltage	V
R_X = base resistor	Ω
R_T = total resistance	Ω
V_S = source voltage	V

Voltage divider networks made up of resistances are often used to obtain a reduced voltage from a power source. Often referred to as the voltage divider equation, this useful tool permits the rapid calculation of the output voltage from such a network. What the equation is actually saying is that the output voltage is equal to the ratio of the base resistor (R_X) to the entire divider resistance (R_T) multiplied by the power source voltage.

In Fig. 11-1, R_X, the base resistor, is 3 Ω, while the total divider resistance is 5 Ω, and the source voltage is 10 V. Dividing the 3 Ω by 5 Ω gives us the factor 0.6, which, when multiplied by 10 V, yields an output voltage of 6 V.

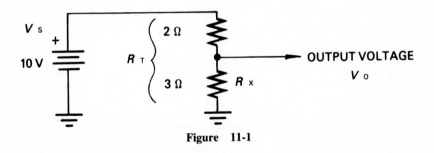

Figure 11-1

Using this equation to solve for R_X instead of V_{OUT} allows you to design your own voltage divider network where:

$$R_X = \frac{V_{OUT} \times R_T}{V_S}$$

This use is shown in Example 11-2. Equation 11 is applicable to series circuits only, that is, where the *current* is the same in all resistors.

EXAMPLE 11-1

Calculate the output voltage V_{OUT} of the voltage divider in Fig. 11-2.

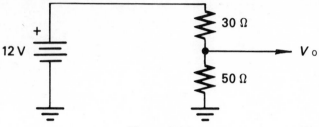

Figure   11-2

First, find the total resistance:

$$R_T = R_1 + R_2$$
$$= 30\ \Omega + 50\ \Omega$$
$$= 80\ \Omega$$

Then calculate the output voltage:

$$V_{OUT} = \frac{R_X}{R_T} \times V_s$$
$$= \frac{50}{80} \times 12\ V$$
$$= 0.625 \times 12\ V$$
$$= 7.5\ V$$

EXAMPLE 11-2

The 90-V battery in Fig. 11-3 drives a 50-kΩ divider network from which a 20-V output voltage is produced. What is the value of each resistor?

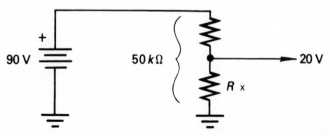

Figure   11-3

First, find the value of the base resistor:

$$R_x = \frac{V_{OUT} \times R_T}{V_s}$$

$$= \frac{20 \times (50 \times 10^3)}{90}$$

$$= \frac{10^6}{90}$$

$$= 11,111 \ \Omega$$

The upper resistor is

$$50,000 - 11,111 = 38,889 \ \Omega$$

EQUATION 12
CURRENT DIVIDER

$$I_{BR} = \frac{R_{OP}}{R_1 + R_2} \times I_T$$

I_{BR} = branch current	A
R_{OP} = opposite resistor	Ω
R_1, R_2 = branch resistors	Ω
I_T = total current	A

Just as the voltage divider equation enables us to find an output voltage as a proportion of the source voltage, Equation 12 allows us to calculate a particular branch current as a proportion of the total current. A ratio of the opposite resistance (i.e., opposite the one in which the current is being calculated) to the total parallel resistance is taken and the result is multiplied by the total current. This equation's practical applicability is fairly limited, since it is far easier to simply use Ohm's law on a particular current path. Its prime importance is as an educational tool that allows us to see that branch currents can be set up on a proportional basis by using the appropriate resistance ratios. For example, find the current through R_1 in Fig. 12-1.

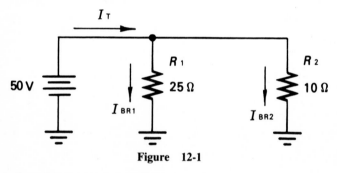

Figure 12-1

First, the total current is found:

$$R_T = \frac{10 \times 25}{10 + 25} = 7.143 \ \Omega$$

$$I_T = \frac{V}{R_T} = \frac{50}{7.143} = 7 \ A$$

Then, the branch current is calculated:

$$I_{BR1} = \frac{10}{25 + 10} \times 7$$

$$= \frac{10}{35} \times 7 = 2 \ A$$

Note that since we are determining the current through R_1, R_2 becomes the opposite branch resistance, R_{OP}. This formula is applicable to parallel circuits only, that is, where the *voltage* across all resistors is the same.

EXAMPLE 12-1

Using Equation 12, determine the current flowing through the lamp in Fig. 12-2.

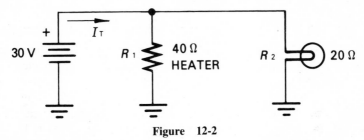

Figure   12-2

First, find the total current if the source is a 30-volt battery.

$$R_T = \frac{R_1 \times R_2}{R_1 + R_2} = \frac{40 \times 20}{40 + 20}$$

$$= 13.33 \, \Omega$$

$$I_T = \frac{V}{R_T} = \frac{30}{13.33} = 2.25 \, A$$

Then the branch current is found.

$$I_{BR} = \frac{R_{OP}}{R_1 + R_2} \times I_T$$

$$= \frac{40}{60} \times 2.25 \, A$$

$$= 0.6667 \times 2.25 \, A$$

$$= 1.5 \, A$$

EXAMPLE 12-2

From the total current obtained in Example 12-1, determine the branch current through the heater.

$$I_{BR} = \frac{R_{OP}}{R_1 + R_2} \times I_T$$

$$= \frac{20}{60} \times 2.25 \, A = 0.3333 \times 2.25 \, A$$

$$= 0.7499 \, A \cong 750 \, mA$$

$$I_L = \frac{V_{TH}}{R_L + R_{TH}}$$

I_L = load current   A
V_{TH} = Thévenin voltage   V
R_{TH} = Thévenin resistance Ω
R_L = load resistance Ω

Thévenizing a circuit allows us to take a complex voltage-resistance network and find an equivalent simple circuit for ease of analysis. All such networks may be reduced to a single voltage and resistance as follows:

(1) To determine the Thévenin resistance, imagine there is no voltage source but instead a simple connecting wire, and calculate the resistance between the load leads as if there were no load connected.
(2) To find the Thévenin voltage, simply imagine the load has been removed and determine the potential between the load terminals by Kirchhoff's voltage law.

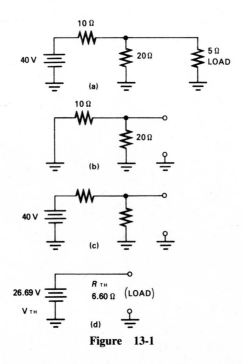

Figure 13-1

Figure 13-1*a* shows the original circuit. ''Shorting'' the source and opening the load, as in Fig. 13-*b*, gives:

$$R_{TH} = \frac{10 \times 20}{10 + 20} = 6.667 \ \Omega$$

Opening the load (Fig. 13-1*c*) gives:

$$V_{TH} = \frac{R_x}{R_T} \times V_s$$

$$= \frac{20}{30} \times 40 = 26.67 \ V$$

resulting in the Thévenin equivalent circuit, shown in Fig. 13-1*d*.

When the Thévenin equivalent circuit has been determined, the 5-Ω load, or any other load, may now be inserted and the voltage across it and the current through it determined. In addition, any number of new loads may be applied and their voltages and currents calculated in rapid succession.

EXAMPLE 13-1

Thévenize the circuit shown in Fig. 13-2 and find the voltage across the load.

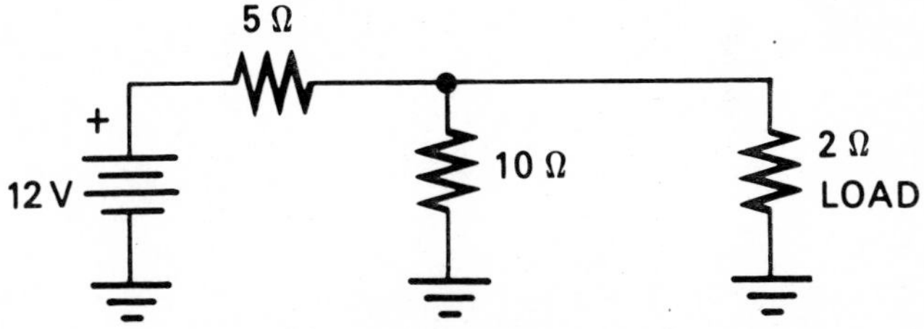

Figure 13-2

$$R_{TH} = \frac{R_1 \times R_2}{R_1 + R_2} = \frac{5 \times 10}{5 + 10} = 3.333 \ \Omega$$

$$V_{TH} = \frac{R_x}{R_T} \times V_s = \frac{10}{15} \times 12 = 8 \ V$$

The Thévenin circuit is shown in Fig. 13-3. The voltage across the load is easily calculated:

$$V_{OUT} = \frac{R_x}{R_T} \times V_s = \frac{2}{5.333} \times 8 = 3 \ V$$

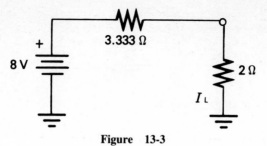

Figure 13-3

EXAMPLE 13-2

Determine the current flowing through the load in Fig. 13-3.

$$I_L = \frac{V_{OUT}}{R_L} = \frac{3}{2} = 1.5 \text{ A}$$

EQUATION 14
WIRE RESISTANCE

$$R = \rho \times \frac{l}{A}$$

R = resistance	Ω
ρ = specific resistance, or resistivity	cmil $\times$ Ω/ft
l = length	ft
A = area	cmil

When long runs of wire are to be made between two pieces of electrical equipment, the actual resistance of the wire itself often becomes an important factor. If the wire resistance between a generator and a lamp, for example, is high enough, some of the voltage will be developed (dropped) across the wire and not the lamp so that the illumination will be poor. Equation 14 may be applied to calculate the exact wire resistance and, from that, the voltage drop on the line. Data from the following two charts are required for use in the resistance equation:

Wire material	ρ
Aluminum	17
Copper	10.4
Nichrome	676

Wire gage	Area (cmil)
10	10,380
12	6,530
14	4,107
18	1,624
20	1,022

What these tables indicate is, first, that copper, for instance, has less resistivity per foot than does aluminum and is therefore a better conductor. Second, 10-gage wire has a larger area and will conduct better than 20-gage wire.

EXAMPLE 14-1

Calculate the resistance of a single 18-gage copper conductor that is 500 feet (ft) long.

From Equation 14:

$$R = \rho \times \frac{l}{A}$$
$$= 10.4 \times \frac{500}{1624}$$
$$= 10.4 \times 0.3078$$
$$= 3.202 \ \Omega$$

EXAMPLE 14-2

Find the voltage across the 75-W, 120-V lamp in Fig. 14-1 when it is 1000 ft from the generator.

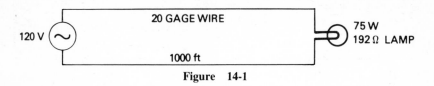

Figure   14-1

First find the resistance of the copper wire:

$$R = \rho \times \frac{l}{A} = 10.4 \times \frac{1000}{1022}$$

$$= 10.17 \ \Omega$$

Since there is 1000 ft of wire in each direction, the total circuit resistance is calculated as follows:

$$R_T = 10.17 \times 2 + 192 = 212.3 \ \Omega$$

Then, using Equation 11, find the voltage:

$$V = \frac{R_x}{R_T} \times V_s$$

$$= \frac{192}{212.3} \times 120$$

$$= 108.5 \ V$$

Therefore, there is only 108.5 V across the lamp, not 120 V.

EQUATION 15
INTERNAL RESISTANCE

$$R_i = \frac{V_{NL} - V_L}{I_L}$$

R_i = internal resistance	Ω
V_{NL} = no-load voltage	V
V_L = load voltage	V
I_L = load current	A

The internal resistance of a power source, such as a battery, will have an effect on the amount of voltage actually applied to the load. If the internal resistance becomes high enough, most of the voltage will be developed across it and not at the load, where it belongs. This becomes very evident in an old or worn-out battery, where the internal resistance has become high and the battery is unable to deliver any appreciable amount of current to the load. Well-designed power supplies or fresh batteries have such low internal resistance that, as the load resistance varies, there is no resistive slack and, as a result, the voltage remains fairly constant. Equation 15 permits the calculation of internal resistance by comparing the loaded and unloaded voltages. If little or no change results in the numerator of the equation, an internal resistance of close to 0 exists. It is necessary to calculate internal resistance in this loaded/nonloaded fashion. It would be impossible to employ an ohmmeter because of the voltages present. Although R_i is actually composed of an infinite number of very small resistances within the power source, we often represent it as a small resistor placed in series with the voltage source as shown in Fig. 15-1.

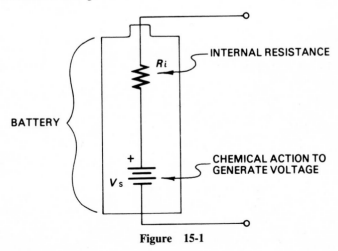

Figure   15-1

EXAMPLE 15-1

What is the internal resistance of the 6-V battery in Fig. 15-2? Assume the voltmeter has a high enough resistance that it does not load the source.

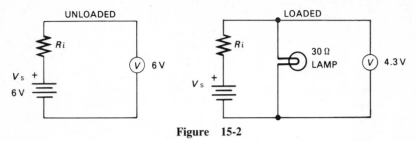

Figure 15-2

First, measure the load voltage to determine the load current:

$$I_L = \frac{V}{R} = \frac{4.3}{30} = 0.1433 \text{ A}$$

Then, using Equation 15, find the internal resistance:

$$R_i = \frac{V_{NL} - V_L}{I_L}$$

$$= \frac{6 - 4.3}{0.1433} = \frac{1.7}{0.1433} = 11.86 \text{ Ω}$$

EXAMPLE 15-2

Use the voltage divider equation (Equation 11) to prove that Example 15-1 is true. (See Fig. 15-3.)

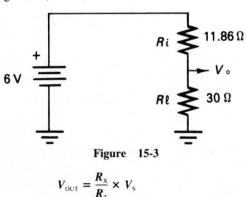

Figure 15-3

$$V_{OUT} = \frac{R_X}{R_T} \times V_s$$

$$= \frac{30}{41.86} \times 6$$

$$= 0.7167 \times 6 = 4.3 \text{ V}$$

EQUATION 16
FLUX DENSITY

$$B = \frac{\phi}{A}$$

B = flux density   G
ϕ = flux lines   Mx
A = area   cm^2

We can compare various magnetic devices, such as permanent magnets and electromagnets, in terms of flux density. This is a way of describing how much magnetic energy is concentrated into a particular portion of the field surrounding a magnet. The maxwell (Mx) itself equals one magnetic field line. When applied to a given section area of the overall field, the number of field lines per unit of area yields flux density (Fig. 16-1). For example, in Fig. 16-2, we can see in end view that 13 field lines pass through an area of 10 cm^2.

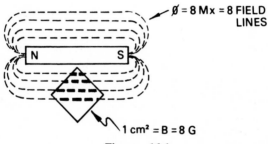

ϕ = 8 Mx = 8 FIELD LINES

1 cm^2 = B = 8 G

Figure   16-1

ϕ = 13 Mx

10 cm^2

Figure   16-2

$$B = \frac{\phi}{A} = \frac{13}{10} = 1.3 \text{ G}$$

Therefore, the flux density becomes 1.3 gauss.

EXAMPLE 16-1

240 Mx passes through a portion of space measuring 40 mm × 30 mm. What is the flux density?

First, determine the area in square centimeters:

$$40\,mm = 4\,cm \quad and \quad 30\,mm = 3\,cm$$
$$3\,cm \times 4\,cm = 12\,cm^2$$

Then, use Equation 16 to solve for flux density:

$$B = \frac{\phi}{A} = \frac{240}{12} = 20\,G$$

EXAMPLE 16-2

If 1 weber (Wb) equals 1 × 10⁸ Mx and 0.0001 Wb passes through a 5-cm² area, what is the flux density?

First, convert the 0.0001 Wb to maxwells:

$$1 \times 10^8 \times 0.0001 = 10,000\,Mx$$

Then, use Equation 16 to solve for flux density:

$$B = \frac{\phi}{A} = \frac{10,000}{5} = 2000\,G$$

Note It would take a permanent magnet weighing several pounds to produce the flux density in Example 16-2.

EQUATION 17
AMPERE-TURNS

$T = NI$	T = ampere-turns
	N = number of turns
	I = current A

Equation 17 is a basic statement from which we can derive two useful formulas. First:

$$N = \frac{T}{I}$$

for determining the number of turns, and second:

$$I = \frac{T}{N}$$

for finding current.

The strength of an electromagnetic field is determined by two factors: the current flowing through the coil and the number of turns in the coil itself. Increasing either will increase the overall magnetizing force, which is also referred to as *magnetomotive force*. From this we can see that a given magnet with a higher value of ampere-turns than another has a greater degree of magnetism. It is also apparent from Equation 17 that a 20-turn electromagnet with a current flow of 2 A produces the same magnetic pull as one with 10 turns and a current flow of 4 A. This is true only if the core material of the electromagnet is not changed.

EXAMPLE 17-1

How much current is necessary to produce a magnetizing force of 300 ampere-turns in an electromagnet with 100 turns of wire?

$$I = \frac{T}{N} = \frac{300}{100} = 3 \text{ A}$$

EXAMPLE 17-2

How many turns are required to produce an NI of 1000 ampere-turns with a current flow of 4 A?

$$N = \frac{T}{I} = \frac{1000}{4} = 250 \text{ turns}$$

EQUATION 18
INDUCTANCE

$$L = \mu \times \frac{N^2 \times A}{l} \times 1.26 \times 10^{-6}$$

L = inductance	H
μ = permeability	
N = number of turns	
A = area	m²
l = length	m

Unlike resistors and capacitors, many inductors (coils) are found without designated values. Equation 18 is useful for determining the unknown inductance if several other factors are known. Note that the area and length of the coil are in metric units. Should you be measuring your particular coil in English units, the following will aid in conversion:

1 inch (in) = 0.0254 meters (m)

The *area* referred to in the equation is taken at the end of the coil, as shown in Fig. 18-1. If diameter d = 12 mm, then radius r = 6 mm and

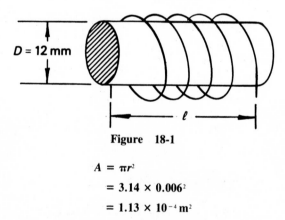

Figure 18-1

$$A = \pi r^2$$
$$= 3.14 \times 0.006^2$$
$$= 1.13 \times 10^{-4} \, m^2$$

The only other term which bears explanation is μ, which is the permeability of the core material. Air has a μ of 1, while the μ of iron is about 400. Thus, if we insert an iron core into our coil, the inductance will be increased greatly. This is due to the fact that ferrous material greatly concentrates the magnetic flux, which in turn increases the inductive effect.

It should also be noted that this equation applies only when the length of the coil is much greater than the diameter (L ≫ D), and when the coil consists of a single layer of close-spaced windings.

EXAMPLE 18-1

Determine the inductance of the coil in Fig. 18-2a.
After first converting to metric (Fig. 18-2b), determine the area:

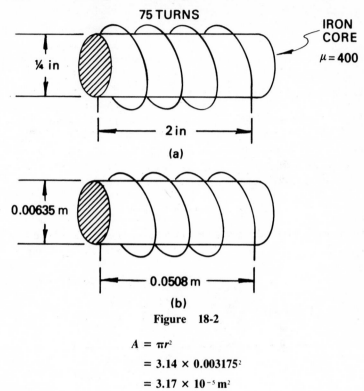

75 TURNS

IRON CORE

$\mu = 400$

¼ in

2 in

(a)

0.00635 m

0.0508 m

(b)

Figure 18-2

$$A = \pi r^2$$
$$= 3.14 \times 0.003175^2$$
$$= 3.17 \times 10^{-5}\,\text{m}^2$$

Then, using Equation 18:

$$L = \mu \times \frac{N^2 \times A}{l} \times (1.26 \times 10^{-6})$$

$$= 400 \times \frac{75^2 \times (3.17 \times 10^{-5})}{0.0508} \times (1.26 \times 10^{-6})$$

$$= 400 \times \frac{0.1783}{0.0508} \times (1.26 \times 10^{-6})$$

$$= \frac{71.32}{0.0508} \times (1.26 \times 10^{-6})$$

$$= 1404 \times (1.26 \times 10^{-6})$$

$$= 1.77 \times 10^{-3}\,\text{H}$$

$$= 1.77\,\text{mH}$$

<table>
<tr><td colspan="2" align="center">EQUATION 19
INDUCTANCE, SERIES</td></tr>
<tr><td>$L_T = L_1 + L_2 + \cdots + L_n$</td><td>$L_T$ = total inductance   H
L_1, L_2 = inductors   H
L_n = last inductor   H</td></tr>
</table>

The total inductance of coils connected in series is simply equal to the sum of the individual inductances. This is to say, two 16-H inductors connected in series will be the equivalent of a single 32-H unit. In using Equation 19, though, you must be careful that the inductors are spaced far enough apart so that their magnetic fields do not interact. Equations 21 and 22 will take this particular interaction into account.

As a practical application, assume that you need the cemf (collapse voltage) to be very high across your inductor and that one of the 10-H chokes that you have on hand will not produce sufficient voltage. By adding another 10-H unit in series with the first, you will double the voltage when the switch is opened.

EXAMPLE 19-1

Find the total inductance in the circuit of Fig. 19-1.

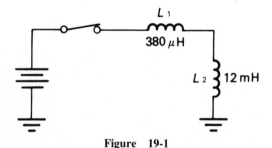

Figure 19-1

First convert all inductances to similar units before adding.

$$12 \text{ mH} = 12{,}000 \text{ μH}$$

Then,

$$L_T = L_1 + L_2$$
$$= 380 \text{ μH} + 12{,}000 \text{ μH}$$
$$= 12{,}380 \text{ μH}$$

EXAMPLE 19-2

Find the total inductance in the circuit of Fig. 19-2.

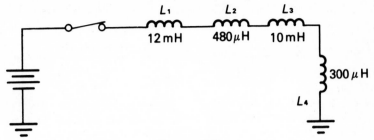

Figure 19-2

$$L_T = L_1 + L_2 + L_3 + L_4$$
$$= 12\,\text{mH} + 0.48\,\text{mH} + 10\,\text{mH} + 0.3\,\text{mH}$$
$$= 22.78\,\text{mH}$$

EQUATION 20	
INDUCTANCE, PARALLEL	

$$L_T = \frac{1}{1/L_1 + 1/L_2 + \cdots + 1/L_n}$$

L_T = total inductance   H
L_1, L_2 = inductances   H
L_n = other inductance   H

Equation 20 is similar to Equation 7 for parallel resistance. Equation 20 states that as more inductors are placed in parallel, the overall inductance will decrease. The sum of all the inductors involved will always be less than the value of the lowest single inductance. For example, if we parallel a 4-H unit with a 6-H one, the total inductance is 2.4 H, which is less than 4.

As with Equation 19, we can apply this parallel equation only if there is no mutual inductance between coils. This occurs when the inductors are too close to each other.

EXAMPLE 20-1

What must the total inductance obviously be less than in the circuit of Fig. 20-1?

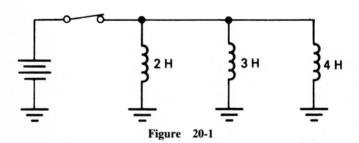

Figure 20-1

The total inductance must be less than 2 H. This is because 2 H is the lowest single value in the circuit and the inductors are in parallel.

EXAMPLE 20-2

Calculate the total inductance in the circuit of Fig. 20-2.

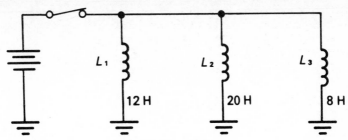

Figure   20-2

$$L_{\text{T}} = \frac{1}{1/L_1 + 1/L_2 + 1/L_3}$$

$$= \frac{1}{1/12 + 1/20 + 1/8}$$

$$= \frac{1}{0.083 + 0.05 + 0.125}$$

$$= \frac{1}{0.258}$$

$$= 3.876 \text{ H}$$

EQUATION 21
MUTUAL INDUCTANCE

$L_m = k \sqrt{L_1 L_2}$	L_m = mutual inductance	H
	k = coefficient of coupling	
	L_1 = first inductance	H
	L_2 = second inductance	H

When a changing current in one coil induces a current in another because of an interaction of their magnetic fields, a mutual inductance affects the circuit. Thus, if a 4-H coil and a 2-H coil are placed in series end to end and the windings of both are going in the same direction, the total inductance will be greater than 6-H.

The actual calculation of mutual inductance is of little use in itself but becomes an integral part of Equation 22. It is for this reason that we pursue it here.

The only term that needs explanation is k, the coefficient of coupling, which is the ratio of the flux from one coil to the total flux from both coils. Two coils wound one directly over the other would be tightly coupled, and in this case k would be about 0.25 on an air core. An iron core *could* raise k to its maximum value of 1. Spacing the coils farther apart is referred to as *loose coupling* and could drive the factor to 0. Placing the inductors at right angles also will reduce k. Note also that L_1 and $_2$ must be converted to henrys before calculating L_m.

EXAMPLE 21-1

The two 380-mH air-core coils in Fig. 21-1 are placed close enough so that their coefficient of coupling k is 0.14. Using Equation 21, find what mutual inductance is produced.

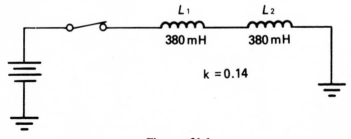

Figure 21-1

$$L_m = k\sqrt{L_1 L_2}$$
$$= 0.14\sqrt{(0.38)(0.38)}$$
$$= 0.14 \times 0.38$$
$$= 0.053 \text{ H} = 53 \text{ mH}$$

EXAMPLE 21-2

Using higher inductances and a higher value of k than in Example 21-1 results in a much greater mutual effect. In Fig. 21-2, two 4.7-H coils are wound together on an iron-core form to produce a k of 0.6. What mutual inductance is produced?

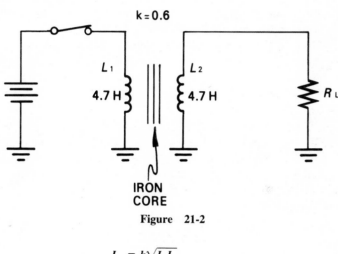

Figure 21-2

$$L_m = k\sqrt{L_1 L_2}$$
$$= 0.6\sqrt{(4.7)(4.7)}$$
$$= 0.6 \times 4.7$$
$$= 2.82 \text{ H} = 2820 \text{ mH}$$

<table>
<tr><td colspan="3" align="center">EQUATION 22
INDUCTANCE, SERIES, WITH L_m</td></tr>
</table>

$L_T = L_1 + L_2 \pm 2L_m$	L_T = total inductance	H
	L_1 = first inductance	H
	L_2 = second inductance	H
	L_m = mutual inductance	H

Equation 22 is used when two coils are placed in series and are close enough to each other to have mutual inductance. Mutual inductance (L_m) may be calculated by Equation 21. In solving for L_T we must first determine whether to add or subtract the factor of $2L_m$. If both coils are wound in the same direction, $2L_m$ is added. Coils wound in this fashion are said to be *series-aiding*. If they are wound in opposite directions, $2L_m$ is subtracted. Coils wound in this manner are called *series-opposing*. Refer to Fig. 22-1.

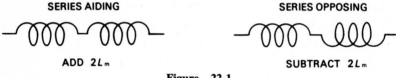

SERIES AIDING   SERIES OPPOSING

ADD $2L_m$ SUBTRACT $2L_m$

Figure 22-1

EXAMPLE 22-1

Determine the total inductance of the circuit in Fig. 22-2.

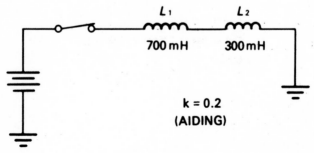

L_1 L_2

700 mH   300 mH

k = 0.2
(AIDING)

Figure 22-2

First determine L_m:

$$L_m = k \sqrt{L_1 L_2}$$
$$= 0.2 \sqrt{(0.7)(0.3)}$$
$$= 0.2 \times 0.458$$
$$= 0.092 \text{ H} = 92 \text{ mH}$$

Since the coils are series-aiding, $2L_m$ is added:

$$L_T = L_1 + L_2 + 2L_m$$
$$= 700 + 300 + 2\,(92)$$
$$= 1184 \text{ mH}$$

EXAMPLE 22-2

Determine the total inductance of the circuit in Fig. 22-3.

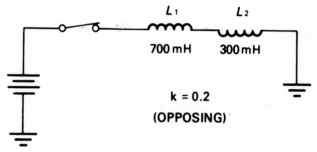

Figure   22-3

From Example 22-1 above, $L_m = 92$ mH. This time, the coils are series-opposing. Thus,

$$L_T = L_1 + L_2 - 2L_m$$
$$= 700 + 300 - 2\,(92) = 1000 - 184$$
$$= 816 \text{ mH}$$

EQUATION 23 CAPACITOR CHARGE		
$Q = C \times V$	Q = charge	C
	C = capacitance	F
	V = voltage	V

One of the prime functions of a capacitor is to store an electric charge. The amount of charge (the number of electrons) that is stored is determined by both the size of the capacitor (in farads) and the amount of applied voltage (in volts). Although the basic unit of capacitance, the farad, is used in the formula, actual capacitors are in the microfarad (μF) or picofarad (pF) size range. To see just how this formula is applied, let us assume that 6 V is applied across two capacitors, one twice as large as the other.

$$Q = C \times V$$
$$= (1 \times 10^{-6}) \times 6$$
$$= 6 \times 10^{-6}$$

$$Q = C \times V$$
$$= (2 \times 10^{-6}) \times 6$$
$$= 12 \times 10^{-6}$$

When the batteries are removed, both capacitors will have a 6-V potential across them; however, the 2-μF capacitor will have twice as much charge and, therefore, will discharge into a given load for twice as long a period.

EXAMPLE 23-1

Determine the amount of charge that is stored in the capacitor of Fig. 23-1.

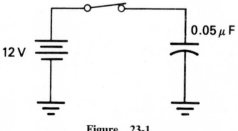

Figure 23-1

From Equation 23:

$$Q = C \times V$$
$$= (0.05 \times 10^{-6}) \times 12$$
$$= 6 \times 10^{-7} \, C$$

EXAMPLE 23-2

Prove that if 12 V is applied to a 0.005-μF capacitor instead of 6 V a greater charge will be stored.

Using Equation 23, find the charge that will be stored when each voltage is applied:

$$Q = C \times V \qquad\qquad Q = C \times V$$
$$= (0.005 \times 10^{-6}) \times 12 \qquad = (0.005 \times 10^{-6}) \times 6$$
$$= 6 \times 10^{-8} \, C \qquad\qquad = 3 \times 10^{-8} \, C$$

Thus, the capacitor with 12 V applied to it will store a charge of 6×10^{-8} C. This is twice the charge the capacitor with 6 V applied to it will store.

EQUATION 24
CAPACITOR VOLTAGE

$$V = \frac{Q}{C}$$

V = voltage   V
Q = charge   C
C = capacitance   F

This form of Equation 23 allows us to calculate the voltage across the capacitor when both the charge and the capacitance are known. The charge may be derived from current and time as in Equation 2 and then substituted in Equation 24. This is advantageous because we must have Q for the equation and instruments for measuring Q are not readily available; current and time, however, are easily measured.

For example, if we have charged a 0.15-μF capacitor for 2 s with a current of 7 μA, we can calculate the charge as follows:

$$Q = I \times T = (7 \times 10^{-6}) \times 2 = 1.4 \times 10^{-5}\,C$$

We can now determine the voltage across the capacitor by using the calculated charge:

$$V = \frac{Q}{C} = \frac{1.4 \times 10^{-5}}{0.15 \times 10^{-6}} = 93.3\,V$$

EXAMPLE 24-1

If the charge stored in a 0.5-μF capacitor is 3×10^{-5} C, what is the voltage across that capacitor? (See Fig. 24-1.)

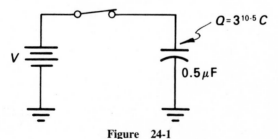

$Q = 3^{10-5}\,C$

$0.5\,\mu F$

Figure 24-1

$$V = \frac{Q}{C} = \frac{3 \times 10^{-5}}{0.5 \times 10^{-6}} = 60\,V$$

EXAMPLE 24-2

A 1-μF capacitor is charged with 12 μA of current for 3 s. What voltage is present across the capacitor? (See Fig. 24-2.)

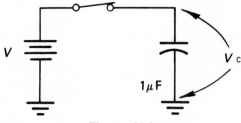

Figure 24-2

First find Q:

$$Q = I \times T = (12 \times 10^{-6}) \times 3 = 3.6 \times 10^{-5} \, C$$

Then use this value of Q to determine the voltage:

$$V = \frac{Q}{C} = \frac{3.6 \times 10^{-5}}{1 \times 10^{-6}} = 36 \, V$$

EQUATION 25
CAPACITANCE, SERIES

$$C_T = \frac{1}{1/C_1 + 1/C_2 + \cdots + 1/C_n}$$

C_T = total capacitance   F
C_1, C_2 = capacitors   F
C_n = other capacitor   F

When totaling series resistances in Equation 5 we simply added them together. Here, with series capacitors, we have just the opposite effect. Capacitors connected in series have an overall circuit effect of less capacitance. The total capacitance must be less than the value of any single capacitor in series. This often proves useful when we do not have on hand the exact value of capacitance we need. For example, if a 0.15-μF capacitor provides twice the capacitance required, placing two in series will produce 0.075 μF.

A beneficial result of placing capacitors in series is that the source voltage divides itself across the units. Thus, applying 300 V to three 0.15-μF capacitors requires each to have a voltage rating of only 100 V. Of course, in actual practice, 150-V ratings would be employed just to be on the safe side.

EXAMPLE 25-1

What must the total capacitance obviously be less than in Fig. 25-1?

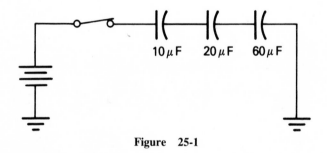

10 μF   20 μF   60 μF

Figure 25-1

Since the capacitors are in series, the total capacitance must be less than 10 μF, that being the lowest value.

EXAMPLE 25-2

Using Equation 25, determine the total series capacitance from point *A* to point *B* in Fig. 25-2.

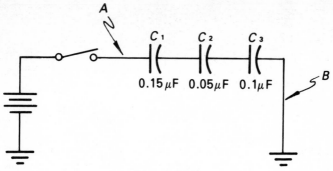

Figure   25-2

$$C_T = \frac{1}{1/C_1 + 1/C_2 + 1/C_3}$$

$$= \frac{1}{1/(0.15 \times 10^{-6}) + 1/(0.05 \times 10^{-6}) + 1/(0.1 \times 10^{-6})}$$

$$= \frac{1}{6.7 \times 10^{-6} + 2 \times 10^7 + 1 \times 10^7}$$

$$= \frac{1}{3.67 \times 10^7}$$

$$= 0.027 \ \mu F$$

EQUATION 26		
CAPACITANCE, PARALLEL		
$C_T = C_1 + C_2 + \cdots + C_n$	C_T = total capacitance	F
	C_1, C_2 = capacitors	F
	C_n = other capacitors	F

When capacitors are connected in parallel, their total capacitance is equal to the sum of the individual capacitances. If, for instance, we are building a power supply that requires a 10,000-μF capacitor and in our stock there is no such value, paralleling two 5000-μF units will give us the 10,000 μF required.

Placing capacitors in parallel does not increase their voltage rating, as the full source voltage is applied across all parallel units. A physical advantage is gained, however, in designing low-profile enclosures. Paralleling many narrower, lower-value units can provide a reduction in height from that needed by a single bulky capacitor.

In checking your answer after a calculation, make certain that it is larger than the largest single value in the network.

EXAMPLE 26-1

Find the total capacitance of the circuit in Fig. 26-1.

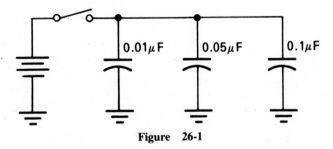

Figure   26-1

From Equation 26,

$$C_T = C_1 + C_2 + C_3$$

$$= 0.01 + 0.05 + 0.1$$

$$= 0.16 \ \mu\text{F}$$

*Microfarads and picofarads are usually used here, as they are common. But be sure only like terms are used.

EXAMPLE 26-2

We have on hand many capacitors with each of the following values: 1 μF, 10 μF, and 20 μF. Our circuit requires a total capacitance of 23 μF. Select the proper combination of capacitors to total 23 μF.

Two combinations are:

$$20 \,\mu F + 1 \,\mu F + 1 \,\mu F + 1 \,\mu F = 23 \,\mu F$$

and $$10 \,\mu F + 10 \,\mu F + 1 \,\mu F + 1 \,\mu F + 1 \,\mu F = 23 \,\mu F$$

While paralleling twenty-three 1-μF capacitors would also be correct, it isn't very practical.

EQUATION 27
RL TIME CONSTANT

$$T = \frac{L}{R}$$

T = time   s
L = inductance   H
R = resistance Ω

In Fig. 27-1, when the switch is closed, the coil will immediately produce a countervoltage, which will, in turn, retard current flow momentarily. With

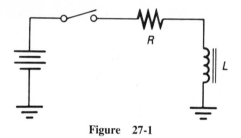

Figure 27-1

Equation 27, you can calculate one time period (or time constant) of this delay. When five of these time periods have passed, the circuit has stabilized. Note that when the resistance is decreased, the entire process is slowed down, as it also is when the inductance is increased. The action is shown graphically in Fig. 27-2.

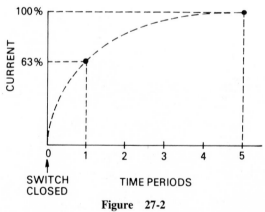

Figure 27-2

The gradually increasing current raises the voltage across the resistor from 0 V at time 0 to full voltage after the fifth time period. After one time constant, the current has reached 63 percent of its full value and therefore the voltage across the resistor has also reached 63 percent of its full value. The voltage across the coil, on the other hand, falls from full voltage at time 0 to 0 V after the fifth time period.

EXAMPLE 27-1

How long does it take the current in the circuit of Fig. 27-3 to reach 63 percent of its total?

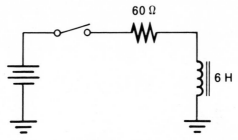

Figure 27-3

Since 63 percent of the full current is reached after the first time period, use Equation 27 to find the time constant.

$$T = \frac{L}{R} = \frac{6}{60} = 0.1 \text{ s}$$

EXAMPLE 27-2

How long will it take for the lamp in Fig. 27-4 to come up to full brilliance after the switch has been closed?

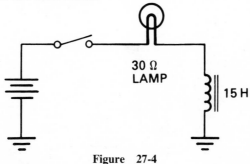

Figure 27-4

First determine the time constant for one time period.

$$T = \frac{L}{R} = \frac{15}{30} = 0.5 \text{ s}$$

The lamp will reach full brilliance after five time periods. Thus,

5 time periods = 5 × 0.5 = 2.5 s

EQUATION 28
RC TIME CONSTANT

$T = RC$	T = time	s
	R = resistance	Ω
	C = capacitance	F

When a voltage is applied to a capacitor, as in Fig. 28-1, current flow is at maximum immediately, but the voltage across the capacitor is not. This

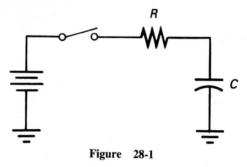

Figure   28-1

voltage will build up to the full source voltage with a speed that is determined by the size of both the resistor and the capacitor. This voltage buildup can be seen graphically in Fig. 28-2.

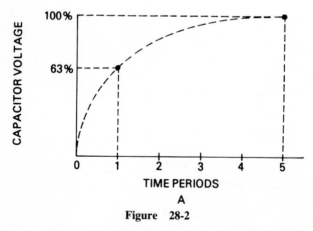

Figure   28-2

Equation 28 will allow us to determine the length of time required for the capacitor to reach 63 percent of full charge. This occurs at the end of the first time period. For a full charge, five time periods are required. Notice that the action of this circuit is opposite to that of the *RL* circuit. While the curve shown in Fig. 28-2 is the same as that shown in Fig. 27-2, it is the *voltage* that rises across a capacitor from 0 V at time 0 to full value after five time periods; the current, on the other hand, falls from its maximum value at time 0 to 0 A after five time periods.

EXAMPLE 28-1

How long is one time period for the circuit of Fig. 28-3?

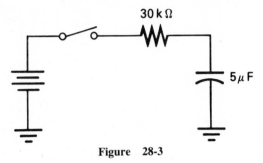

Figure   28-3

From Equation 28,

$$T = RC = 30,000 \times (5 \times 10^{-6}) = 0.15\,s$$

EXAMPLE 28-2

How long will it take for the capacitor in Fig. 28-4 to reach the *full* battery voltage of 9 V?

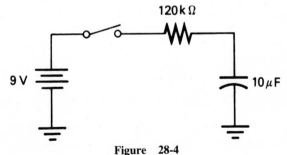

Figure   28-4

First determine the length of one time period.

$$T = RC = (120 \times 10^{3}) \times (10 \times 10^{-6})\,1.2\,s$$

The capacitor will be fully charged after five periods. Thus,

$$5T = 6\,s$$

EQUATION 29
SHUNT RESISTANCE

$$R_{SH} = \frac{R_m \times I_m}{I_T - I_m}$$

R_{SH} = shunt resistance	Ω
R_m = meter resistance	Ω
I_m = meter current	A
I_T = total circuit current	A

The majority of panel meters are of the high-sensitivity, low-current type. If we were to use one of these to measure high currents, the meter movement would be damaged. To prevent this, a shunt resistor is connected in parallel with the meter terminals so that most of the high current passes through the shunt. The small remainder flows through the meter.

Equation 29 permits the ohmic calculation for the proper shunt resistance. It states that when we divide $R_m \times I_m$—which yields the voltage across the meter—by the current which must pass through the shunt ($I_T - I_m$) we will derive the shunt resistance itself. After you have made the ohmic calculation you should calculate the amount of power that will be dissipated by the shunt so that you can select an appropriate-sized unit. Failure to do so could cause the shunt to burn out and the meter, in turn, to be damaged.

EXAMPLE 29-1

A meter with a coil resistance of 1500 Ω that normally reads full scale at 1 mA is to be used in Fig. 29-1 to indicate 2 A at full scale. What should the ohmic value of the shunt be?

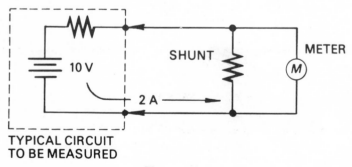

Figure 29-1

From Equation 29,

$$R_{SH} = \frac{R_m \times I_m}{I_T - I_m}$$

$$= \frac{1500 \times 0.001}{2 - 0.001}$$

$$= \frac{1.5}{1.999} = 0.75 \ \Omega$$

EXAMPLE 29-2

If $R_m = 1000 \ \Omega$ and $I_m = 5$ mA, calculate R_{SH} for a full-scale current of 500 mA. Also determine the shunt voltage.

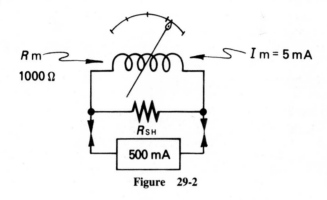

Figure   29-2

From Equation 29,

$$R_{SH} = \frac{R_m \times I_m}{I_T - I_m} = \frac{1000 \times 0.005}{0.5 - 0.005}$$

$$= \frac{5}{0.495} = 10.1 \ \Omega$$

To find the power dissipation of the shunt, use the current through the shunt only:

$$P = I^2 R = 0.495^2 \times 10.1 = 2.47 \ W$$

To be safe, double this result and use a 5-W resistor.

EQUATION 30
METER SENSITIVITY

$$S = \frac{1}{I_m}$$

S = sensitivity	Ω/V
I_m = meter current at full scale	A

Equation 30 indicates the sensitivity of a coil-type meter movement. A knowledge of a meter's ohms per volt (Ω/V) rating is useful for several reasons. First, the higher the value, the more sensitive the meter; we want to use a sensitive meter, say 10,000 Ω/V or higher, in a high-resistance, low-current measurement where the driving signal is weak. (Since this type is more expensive than a less sensitive meter, we would not use it in a strictly high-current application.) Second, knowing the Ω/V value allows us to quickly determine the series resistor required to make the meter function as a voltmeter, as will be explained under Equation 31.

EXAMPLE 30-1

What is the sensitivity of a 1-mA meter movement?
From Equation 30,

$$S = \frac{1}{I_m} = \frac{1}{0.001} = 1000 \ \Omega/V$$

EXAMPLE 30-2

What is the sensitivity of a 100-μA meter movement?

$$S = \frac{1}{I_m} = \frac{1}{100 \times 10^{-6}} = 10,000 \ \Omega/V$$

EQUATION 31
VOLTMETER RESISTANCE

$$R_S \cong V \times S$$

R_S = series resistance	Ω
V = full-scale voltage	V
S = meter sensitivity	Ω/V

When you want to convert a current meter to a voltmeter you must place a resistance in series with the meter itself. Equation 31 will determine the ohmic value of that resistance to a good approximation. The actual function of the series resistor is to prevent a high current from reaching the meter and to drop most of the voltage that would otherwise be dropped across the movement. The result of this series resistance is that the voltmeter has a much higher input resistance than a current meter with its low-resistance shunt. To determine the required value of series resistance, multiply the meter's known sensitivity (Equation 30) by the amount of source voltage that you wish the needle to indicate at full scale. The value you derive will be fairly accurate. For more precise calculations, you have to consider the ohmic value of the meter coil.

EXAMPLE 31-1

We wish to convert a 50-μA meter (with a 3000-Ω coil) into one which will indicate 5 V at full scale. What is the value of the series resistor? (See Fig. 31-1.)

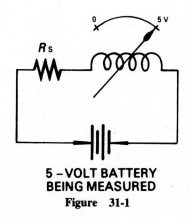

5 – VOLT BATTERY
BEING MEASURED
Figure   31-1

First, solve for the meter sensitivity:

$$S = \frac{1 \text{ V}}{I_m} = \frac{1}{50 \times 10^{-6}} = 20{,}000 \ \Omega/\text{V}$$

Then solve for the series resistance:

$$R_s = V \times S = 5 \times 20{,}000 = 100{,}000$$

EXAMPLE 31-2

Recalculate Example 31-2 taking the 3000-Ω coil resistance into account for an even more precise answer.

$$\textbf{Voltage for full-scale needle deflection} = I \times R$$
$$= (50 \times 10^{-6}) \times 3000$$
$$= 0.15 \text{ V}$$

Thus the voltage across R_S is

$$5 \text{ V} - 0.15 \text{ V} = 4.85 \text{ V}$$

and

$$R_s = \frac{V}{I} = \frac{4.85}{50 \times 10^{-6}} = 97{,}000 \ \Omega$$

AC CIRCUITS

EQUATION 32
INSTANTANEOUS AC VOLTAGE

$$v = V_{MAX} \times \sin\theta$$

v = instantaneous voltage V

V_{MAX} = maximum voltage V

θ = generator angle

It is apparent from the AC waveform of Fig. 32-1 that the maximum voltage is sustained only briefly before falling. In this ever-changing wave

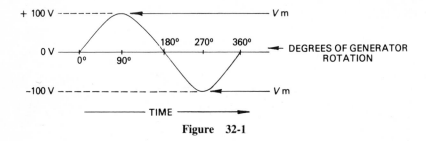

Figure   32-1

it is often necessary to determine the exact voltage at a given instant, for example, at the 45° point. From the diagram you can see that this voltage is somewhere between 0 and 100 V. By employing Equation 32 and using the sine of 45°, you may calculate the exact voltage. To verify the performance of the equation, calculate v at the 90° and 270° points. This should yield the maximum itself. Thus,

$$v = V_{MAX} \times \sin\theta \qquad\qquad v = V_{MAX} \times \sin\theta$$

$$= 100 \times \sin 90° \qquad\qquad = 100 \times \sin 270°$$

$$= 100 \times 1 \qquad\qquad = 100 \times (-1)$$

$$= 100 \text{ V} \qquad\qquad = -100 \text{ V}$$

These are the maximum points, and calculations at any other angles will result in lesser voltage values. For example, at 180°, $v = 0$ volts. Note that V_{MAX} and V_{PEAK} have the same meaning.

EXAMPLE 32-1

Find the instantaneous voltage of the sine wave in Fig. 32-2 when it is at the 200° point. The peak voltage is 30 V.

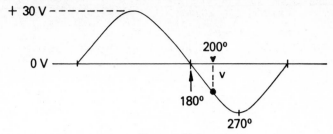

Figure   32-2

$$v = V_{MAX} \times \sin \theta$$
$$= 30 \times \sin 200°$$
$$= 30 \times (-0.342)$$
$$= -10.26 \text{ V}$$

EXAMPLE 32-2

Prove that the instantaneous voltages of a sine wave at 90° and 270° are equal to the positive and negative maximum voltages. Assume V_{MAX} to be 100 V.

$$v = V_{MAX} \times \sin \theta \qquad v = V_{MAX} \times \sin \theta$$
$$= 100 \times \sin 90° \qquad = 100 \times \sin 270°$$
$$= 100 \times 1 \qquad = 100 \times (-1)$$
$$= +100 \text{ V} \qquad = -100 \text{ V}$$

EQUATION 33
AVERAGE VOLTAGE—SINGLE ALTERNATION

$V_{AV} = 0.637 \times V_{MAX}$	V_{AV} = average voltage V of a single alternation 0.637 = a constant V_{MAX} = maximum voltage V

If we were to average all the sine values for a single alternation, the result would be 0.637 times the peak value. For example, sin 30° = 0.5; sin 45° = 0.707, sin 80° = 0.9848. Adding 0.5 and 0.707 and 0.9848 gives 2.1918, which when divided by 3 equals 0.7. If all the angles were used, the more accurate answer of 0.637 would be obtained. Multiplying this factor by a 100-V peak wave yields 63.7 V. Since the wave is constantly increasing and decreasing in amplitude, this calculation allows us to determine the average voltage during this change.

Figure 33-1 shows graphically that the 0.637 level is a true average of areas of voltage (X) and nonvoltage (Y). Here the alternations have all been placed above the zero line for analysis. The average is 0.637 because for this level area X equals area Y.

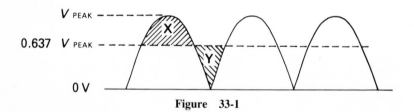

Figure 33-1

EXAMPLE 33-1

What is the average voltage of a sine wave with a peak voltage of 350 V? From Equation 33,

$$V_{AV} = 0.637 \times V_{MAX} = 0.637 \times 350 = 223 \text{ V}$$

EXAMPLE 33-2

What is the average voltage of the 300-V peak-to-peak wave shown in Fig. 33-2?

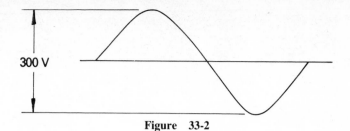

Figure   33-2

First, determine the peak voltage V_{PEAK} from the peak-to-peak value given:

$$V_{\text{PEAK}} = \frac{V_{\text{p}-\text{p}}}{2} = \frac{300}{2} = 150 \text{ V}$$

Then determine V_{AV}.

$$V_{\text{AV}} = 0.637 \times V_{\text{MAX}} = 0.637 \times 150 = 95.5 \text{ V}$$

EQUATION 34
RMS VOLTAGE

$$V_{RMS} = 0.707 \times V_{MAX}$$

V_{RMS} = rms voltage   V
0.707 = a constant
V_{MAX} = maximum voltage   V

When an AC voltage is applied to a resistive load, such as a heater or lamp, we are interested in the power consumed in watts. If peak or average values are used in this calculation, the true power cannot be determined. A value referred to as rms (root mean square), which is derived from the square root of the average of the square of all the sines, when applied to power calculations will yield true power values. The rms voltage value produces the same heat in a load as the DC voltage.

If the sine is taken of all angles from 0 through 180° and these values are then squared and the square root taken of the total divided by the number of calculations, the result will be 0.707. Thus, to find the rms voltage of a 100-V_{PEAK} wave, multiply 0.707 by 100; the result will be 70.7 V rms.

All commercial transformers are rated in rms effective voltages.

EXAMPLE 34-1

If we measure an AC sine wave that has a peak value of 24 V on an oscilloscope, what is the rms value?

$$V_{RMS} = V_{MAX} \times 0.707 = 24 \times 0.707 = 17 \text{ V}$$

EXAMPLE 34-2

A 40-V p − p sine wave is applied to a 200-Ω resistor. How much power is dissipated?

First determine the peak voltage:

$$V_{PEAK} = \frac{V_{p-p}}{2} = \frac{40}{2} = 20 \text{ V}$$

Then calculate the rms voltage:

$$V_{RMS} = V_{MAX} \times 0.707 = 20 \times 0.707 = 14.14 \text{ V}$$

Finally, determine the power using the rms voltage:

$$P = \frac{V^2}{R} = \frac{14.14^2}{200} = 1 \text{ W}$$

EQUATION 35
MAXIMUM VOLTAGE

$$V_{MAX} = 1.414 \times V_{RMS}$$

V_{MAX} = maximum voltage   V
1.414 = a constant
V_{RMS} = rms voltage   V

Equation 34 shows that multiplying the maximum voltage by 0.707 results in an rms value. Equation 35 reverses the process and yields V_{MAX} from the rms voltage. Because the inverse of 1.414 is 0.707, it is also possible to obtain V_{MAX} (the peak voltage) by dividing the rms value by 0.707. Both approaches will give the same answer.

The importance of calculating the maximum voltage from the rms value becomes apparent when we calculate the output level from a power supply. The transformer that supplies alternating current to the supply is listed in rms values by the manufacturer, but the supply produces peak values. With this knowledge it is possible to select the proper transformer for the correct output voltage. Again, remember that the peak value is sustained for only an instant, while the rms value is a sustained average working voltage.

EXAMPLE 35-1

If the rms value of a sine wave is 53 V, what is the maximum value? From Equation 35,

$$V_{MAX} = 1.414 \times V_{RMS} = 1.414 \times 53 = 75 \text{ V}$$

EXAMPLE 35-2

Apply the alternate method for finding V_{MAX} to Example 35-1.

$$V_{MAX} = \frac{V_{RMS}}{0.707} = \frac{53}{0.707} = 75 \text{ V}$$

EQUATION 36
FREQUENCY

$$f = \frac{1}{T}$$

f = frequency	Hz
T = time of one cycle (period)	s

Equations 32 through 35 were all concerned with various properties of the amplitude of the AC wave. Equation 36, however, deals with frequency, or rate of occurrence, of the wave. Some years back, frequency was measured in *cycles per second*. The present term is *hertz* (Hz). The older term exactly defines frequency by stating that the number of cycles that occurs every second is the frequency. When a simple two-pole AC generator makes a single revolution, one wave is produced. If the rotational speed of the generator is now increased, each wave will be produced in a shorter period of time. Therefore, the frequency of the wave is greater when the period of each cycle is less. For example, according to Equation 36 a single cycle of 60-Hz power line voltage occurs in 16.7 ms, whereas the time of a 5-kHz audio wave is only 0.2 ms.

EXAMPLE 36-1

Determine the frequency of the wave of Fig. 36-1.

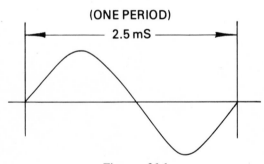

Figure 36-1

From Equation 36,

$$f = \frac{1}{T} = \frac{1}{2.5 \times 10^{-3}} = 400 \text{ Hz}$$

This means that 400 cycles occur every second.

EXAMPLE 36-2

Determine the frequency of the wave of Fig. 36-2.

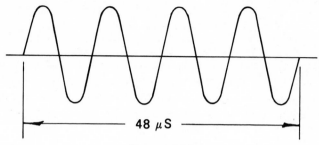

Figure 36-2

The time of the *four cycles* shown is 48 μs; therefore, the period of a single cycle is 12 μs. Then,

$$f = \frac{1}{T} = \frac{1}{12 \times 10^{-6}} = 83.3\,\text{kHz}$$

<table>
<tr><td colspan="2" align="center">**EQUATION 37**
PERIOD</td></tr>
</table>

$$T = \frac{1}{f}$$

T = time of one cycle (period)   s
f = frequency   Hz

If in Equation 36 frequency was equal to the inverse of the time, then, similarly, time is equal to the inverse of the frequency, as shown in Equation 37. This equation states that the period of each cycle decreases as the frequency increases and, conversely, the period increases as frequency decreases. If three successively higher frequencies are compared, we see that the period becomes increasingly shorter.

A + **200 Hz:** $T = \dfrac{1}{200} = 5\,\text{ms}$

A + **1000 Hz:** $T = \dfrac{1}{1000} = 1\,\text{ms}$

A + **20,000 Hz:** $T = \dfrac{1}{2 \times 10^4} = 0.05\,\text{ms}$

At the higher frequencies, the waves are more compressed and the period becomes shorter.

EXAMPLE 37-1

From basic observation, which of the waves in Fig. 37-1 has the shortest period?

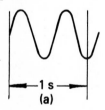
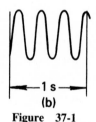
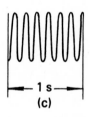

Figure   37-1

Waveform C is the shortest because the greatest number of waves occurs during the same time span.

EXAMPLE 37-2

What is the period of one cycle of a 27-MHz wave?
From Equation 37,

$$T = \frac{1}{f} = \frac{1}{27 \times 10^6} = 0.037\,\mu\text{s}$$

<table>
<tr><td colspan="2" align="center">**EQUATION 38**
GENERATOR FREQUENCY</td></tr>
</table>

$f = \dfrac{P \times S}{120}$	f = frequency Hz P = number of poles S = speed r/min

The basic two-pole generator shown in Fig. 38-1 will produce a sine wave when the armature is rotated. The faster the speed, the higher the output frequency of the wave. It can be seen by examination of Equation 38 that if the armature is rotated at 3600 r/min the output frequency will be 60 Hz. 3600 r/min equals 60 r/s, and if one revolution gives one cycle, then 60 r/s yields 60 Hz.

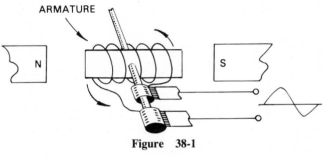

Figure 38-1

$$f = \frac{P \times S}{120} = \frac{2 \times 3600}{120} = 60 \text{ Hz}$$

Equation 38 states that if either the number of poles or the rotational speed is increased, the output frequency will also increase. Decreasing either will decrease the frequency. Should 60 Hz be desired from a lower-speed generator, four poles at 1800 r/min would still result in 60 Hz.

EXAMPLE 38-1

Calculate a generator's output frequency if it has 16 poles and rotates at 1400 r/min.
From Equation 38,

$$f = \frac{P \times S}{120} = \frac{16 \times 1400}{120} = 186.7 \text{ Hz}$$

EXAMPLE 38-2

Double the number of poles in Example 38-1 and recalculate the frequency.

$$f = \frac{P \times S}{120} = \frac{32 \times 1400}{120} = 373.3 \text{ Hz}$$

EQUATION 39
SYNCHRONOUS MOTOR SPEED

$$S = \frac{f \times 120}{P}$$

S = synchronous speed   r/min
f = frequency   Hz
P = number of poles

The speed at which a motor's rotating field turns is called *synchronous speed*. The actual force that turns the motor's shaft is a result of the push and pull forces between the poles of the armature and the surrounding field. Therefore, by increasing the applied frequency, you can increase the rapidity of the push-pull action and thus the motor speed. Conversely, by increasing the number of poles, you will reduce the speed. This reduction results from the dispersion of each cycle of applied voltage in a smaller rotational arc. (The smaller arcs result from the reduced distances between poles due to their greater number.)

The synchronous speed that is calculated by Equation 39 is at a no-load state. If a mechanical load (such as friction) is applied to the shaft of the motor, the rotational speed will decrease because of the resulting slippage within the internal fields.

EXAMPLE 39-1

What is the synchronous speed of a 32-pole motor that is driven with a 50-Hz signal?
From Equation 39,

$$S = \frac{f \times 120}{P} = \frac{50 \times 120}{32} = \frac{6000}{32} = 187.5 \text{ r/min}$$

EXAMPLE 39-2

What is the synchronous speed of a four-pole motor that is driven with a 50-Hz signal?

$$S = \frac{f \times 120}{P} = \frac{50 \times 120}{4} = \frac{6000}{4} = 1500 \text{ r/min}$$

EQUATION 40
WAVELENGTH

$$\lambda = \frac{3 \times 10^8}{f}$$

λ = wavelength   m
3×10^8 = a constant   m/s
f = frequency   Hz

Electronic technicians and engineers often refer to a sine wave in terms of wavelength. The length that is referred to is that of one cycle of the wave. This is to say that one cycle actually has a physical length that can be defined in meters, feet, yards, or other units of length. Equation 40 gives the length in meters; this may then be converted to other linear units if desired. The constant 3×10^8 is the velocity (speed) of any electromagnetic wave in free space given in meters per second. Note that by increasing the frequency you will shorten the wavelength.

One of the practical applications of wavelength measurements is in the design of antennas which selectively capture a signal of a specific frequency. The metallic rods of the antenna are cut to some exact proportion of the wavelength to be received. The term *wavelength* is also used by radio operators to refer to the particular band in which they are communicating in terms of meters.

At 7 MHz,

$$\lambda = \frac{3 \times 10^8}{7 \times 10^6} = 43 \text{ m}$$

The operator would refer to this as the 40-m band.

EXAMPLE 40-1

What is the wavelength of the wave of Fig. 40-1?

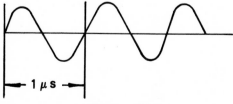

Figure   40-1

First the frequency must be determined from the period:

$$f = \frac{1}{T} = \frac{1}{1 \times 10^{-6}} = 1\,\text{MHz}$$

Now the wavelength can be calculated:

$$\lambda = \frac{3 \times 10^8}{f} = \frac{3 \times 10^8}{1 \times 10^6} = 300\,\text{m}$$

EXAMPLE 40-2

The citizens band is located in what meter band? In other words, what is the wavelength of a 27-MHz signal?

$$\lambda = \frac{3 \times 10^8}{f} = \frac{3 \times 10^8}{27 \times 10^6} = 11.1\,\text{m}$$

The citizens band is in the 11-m band.

EQUATION 41
THREE-PHASE (WYE OR STAR) GENERATOR

$$V_Y = \sqrt{3} \times V$$

V_Y = 3-ϕ voltage   V

V = 1-ϕ voltage   V

$\sqrt{3}$ = a constant

Figure 41-1 depicts a three-phase (3-ϕ) generator connected to a three-phase load and a small single-phase lamp. The load in such a system would most likely be a 3-ϕ industrial motor. The 3-ϕ system permits the use of smaller-gage wire and allows more efficient motor performance over that of a single-phase system. Note also that lamps and similar devices in the area of the motor can pick off 120 V single-phase for their operation from the neutral wire and one coil.

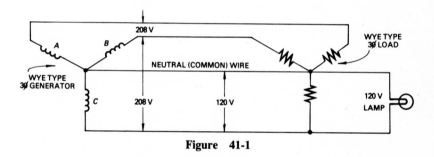

Figure   41-1

By inspection of Fig. 41-1 it can be seen that each leg of the generator is producing 120 V. (Follow the lamp leads back to the generator.) To calculate the voltage across any two legs, multiply the square root of 3 (which is 1.732) by the voltage produced by a single coil. In this case, 120 × 1.732 = 208 V. The full 3-ϕ signal produced appears in Fig. 41-2.

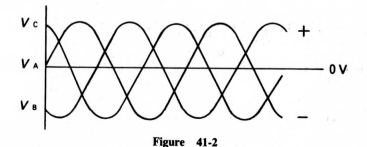

Figure   41-2

EXAMPLE 41-1

The voltage measured in Fig. 41-3 between the neutral wire and the end of one generator coil is 300 V. What is the voltage across a two-coil output?

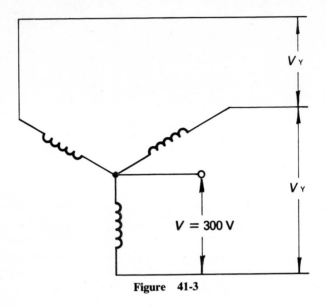

Figure   41-3

From Equation 41,

$$V_Y = \sqrt{3} \times V = 1.732 \times 300 = 520 \text{ V}$$

$$\frac{N_P}{N_S} = \frac{V_P}{V_S}$$

N_P = number of primary turns

N_S = number of secondary turns

V_P = voltage on primary   V

V_S = voltage on secondary   V

A transformer is capable of doing one of the following with AC voltage:

(1) Stepping it up.
(2) Transferring equal voltages from primary to secondary. This is for isolation use only.
(3) Stepping it down.

The ratio of the number of turns on the primary winding to the number on the secondary is the *turns ratio*. For instance, a transformer with 200 turns on the primary and 5 on the secondary would have a step-down turns ratio of 200 to 5 or 40 to 1, which is written 40:1.

Equation 42 shows that the turns ratio is directly proportional to the voltage ratio. Thus, if the step-down ratio is 40:1, the voltage introduced into the primary will be reduced 40 times. In a similar manner, if the turns ratio were 1:7, this would indicate a step-up application in which any voltage introduced into the primary would appear on the secondary 7 times as great. As will be seen in Examples 42-1 and 42-2, any one of the four factors in the ratio may be found by cross multiplying and solving for the unknown quantity.

EXAMPLE 42-1

If a transformer reduces 120 V to 12 V, what is the turns ratio?
Since the voltage is stepped down 10 times and the voltage ratio is the same as the turns ratio, then the ratio is 10:1.

EXAMPLE 42-2

The transformer in Fig. 42-1 has a turns ratio of 17:1* and 240 V is applied to the primary. What will the secondary voltage be?

---

*Assume that this transformer has N_P = 340 turns and N_S = 20 turns. These figures could have been inserted into the equation in place of the 17:1 ratio.

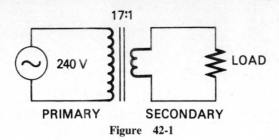

17:1

240 V · LOAD

PRIMARY   SECONDARY

Figure 42-1

From Equation 42,

$$\frac{N_P}{N_S} = \frac{V_P}{V_S}$$

$$\frac{17}{1} = \frac{240}{V_S}$$

$$17\,V_S = 240$$

$$V_S = 14.1\,\text{V}$$

<table>
<thead>
<tr><th colspan="2">EQUATION 43
TRANSFORMER V × I RATIO</th></tr>
</thead>
<tbody>
<tr>
<td>$$\frac{V_P}{V_S} = \frac{I_S}{I_P}$$</td>
<td>V_P = primary voltage   V
V_S = secondary voltage V
I_S = secondary current A
I_P = primary current   A</td>
</tr>
</tbody>
</table>

Most laminated iron-core power transformers have an efficiency rating of about 95 percent. This is to say that the power supplied to the primary results in approximately equal power in the secondary circuit. Since power is the product of voltage and current, if one increases, the other must decrease. Equation 43 states that if the voltage ratio goes up, then the current ratio must decrease. In other words, since $P_{IN} - P_{OUT}$, then $V_P \times I_P = V_S \times I_S$. For example:

Primary   **Secondary**
3 V × 2 A = 6 W   2 V × 3 A = 6 W

Note that as the voltage is stepped down from 3 to 2 V, the current increases from 2 to 3 A, so that P_{OUT} still equals P_{IN}. Heavier *currents* require more massive wire windings; and if the transformer in the example above were examined, we would find a heavier-gage wire on the secondary.

Examples 43-1 and 43-2 will show how to calculate any one of the four quantities if the other three are known by using cross multiplication.

EXAMPLE 43-1

In Fig. 43-1, what current is flowing in the primary?

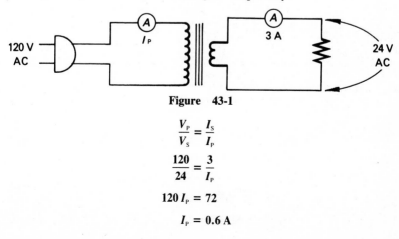

Figure 43-1

$$\frac{V_P}{V_S} = \frac{I_S}{I_P}$$

$$\frac{120}{24} = \frac{3}{I_P}$$

$$120\, I_P = 72$$

$$I_P = 0.6\text{ A}$$

Proof: $P_{IN} = P_{OUT}$

$$120 \times 6 = 24 \times 3$$
$$72\text{ W} = 72\text{ W}$$

EXAMPLE 43-2

What is the secondary voltage in Fig. 43-2?

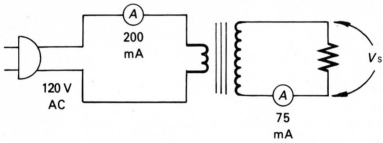

Figure   43-2

$$\frac{V_P}{V_S} = \frac{I_S}{I_P}$$

$$\frac{120}{V_S} = \frac{0.075}{0.2}$$

$$0.075\, V_S = 24$$

$$V_S = 320\text{ V}$$

Proof: $P_{IN} = P_{OUT}$

$$120 \times 0.2 = 320 \times 0.075$$
$$24\text{ W} = 24\text{ W}$$

EQUATION 44	
INDUCTIVE REACTANCE	

$$X_L = 2\pi f L$$

or

$$X_L = \omega L*$$

X_L = inductive reactance Ω

2π = a constant

f = frequency   Hz

L = inductance   H

When a DC or an AC signal is applied to a resistor, the signal *source* "sees" that resistance as a fixed value. If the resistor were a 220-Ω unit (color code red, red, brown), both a DC battery and a 400-Hz AC generator would be working into that 220-Ω load, nothing more, nothing less. Inductors, on the other hand, "react" to a changing AC signal and *present* an ohmic value that is proportional to the frequency of the signal. The ohmic value is referred to as *inductive reactance*.

From Equation 44 it is apparent that if either the frequency goes up or the value of the inductance increases, then the inductive reactance will also increase. Thus, if a 50-mH coil is alternately fed 5 kHz, 10 kHz, and then 15 kHz signals, it will first appear as a 1570-Ω resistance, then a 3140-Ω resistance, and then a 4710-Ω resistance.

The inductance (L) of a coil is a fixed value which does not change. For example, if you purchase a 50-mH coil, this value never changes, but its reactance can, depending on the frequency of the current applied to it.

EXAMPLE 44-1

Determine the inductive reactance of the inductor in Fig. 44-1.

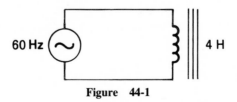

Figure   44-1

From Equation 44,

$$X_L = 2\pi f L = 6.28 \times 60 \times 4 = 1507.2 \ \Omega$$

*The term ω equals $2\pi f$ and is called *angular velocity*.

EXAMPLE 44-2

A 240-V generator can operate at either 60 Hz or 440 Hz. At which frequency will the lamp in Fig. 44-2 be the brightest with the inductor shown in the line?

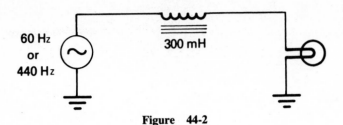

Figure 44-2

60 Hz	440 Hz
$X_L = 2\pi fL$	$X_L = 2\pi fL$
$= 6.28 \times 60 \times 0.3$	$= 6.28 \times 440 \times 0.3$
$= 113 \ \Omega$	$= 829 \ \Omega$

Since at 60 Hz the inductive reactance of the coil is less than at 440 Hz, it will pass a greater amount of current to the lamp, which will be brighter at 60 Hz than at 440 Hz.

EQUATION 45
INDUCTOR Q

$$Q = \frac{X_L}{R}$$

Q = figure of merit

X_L = inductive reactance $\quad \Omega$

R = DC coil resistance $\quad \Omega$

The term Q is often referred to as the *quality* of an inductor. It is the ratio of coil reactance to DC ohmic resistance. To calculate this factor, an ohmic reading is first taken with a multimeter. Then, X_L is calculated by $2\pi fL$ as in Equation 44. The DC resistance reading is then divided into $2\pi fL$. Inductors with a Q factor less than 10 are said to have a poor Q. This means an excessive amount of wire was wound in order to achieve the required inductance or that the wrong core material was used. Coils with poor Q not only complicate AC calculations but also provide poor filtering when used in audio- and radio-frequency circuits. The excessive amount of DC resistance produced by the wire tends to diminish the reactive effect so that inductive action is sluggish.

As frequency increases, inductive reactance also increases. Therefore the Q of any inductor will be greater at higher frequencies.

EXAMPLE 45-1

Calculate the Q of a 5-mH coil at 15 kHz. The DC resistance is 47 Ω.

$$Q = \frac{X_L}{R} = \frac{2\pi fL}{R}$$

$$= \frac{6.28 \times (15 \times 10^3) \times 0.005}{47}$$

$$= \frac{471}{47} = 10$$

EXAMPLE 45-2

Recalculate Example 45-2 at 100 kHz.

$$Q = \frac{X_L}{R} = \frac{2\pi fL}{R}$$

$$= \frac{6.28 \times (100 \times 10^3) \times 0.005}{47}$$

$$= \frac{3140}{47} = 66.8$$

EQUATION 46
CAPACITIVE REACTANCE

$$X_C = \frac{1}{2\pi fC}$$

X_C = capacitive reactance	Ω
2π = a constant	
f = frequency	Hz
C = capacitance	F

Like inductors, capacitors react to AC signals and present an ohmic value to the signal source. When you examine Equation 46, you can see that if either the frequency or the capacitance is increased, the capacitive reactance will decrease. For instance, the ohmic values for a 0.05-μf capacitor at 500 Hz, 10 kHz, and 3 MHz, respectively, are 6369 Ω, 319 Ω, and 1 Ω. Thus, the capacitor reacts to different frequencies and appears as a different resistance to each.

It is now apparent that we must consider the exact operating frequency when dealing with reactive devices such as capacitors. Ohmic values thus derived can be used to calculate currents and voltage drops in the circuit being examined.

Capacitive reactance, although in ohms like inductive reactance, is 180° out of phase with its counterpart; if both are in a series circuit together, their respective values must be subtracted from each other.

EXAMPLE 46-1

Find the capacitive reactance of the 10-μF capacitor in Fig. 46-1 at 600 Hz.

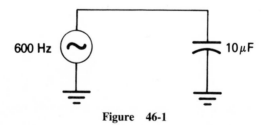

Figure 46-1

From Equation 46,

$$X_C = \frac{1}{2\pi fC} = \frac{1}{6.28 \times 600 \times (10 \times 10^{-6})}$$
$$= 26.5 \ \Omega$$

EXAMPLE 46-2

Recalculate the capacitive reactance of the 10-μF capacitor in Example 46-1 at 50 Hz.

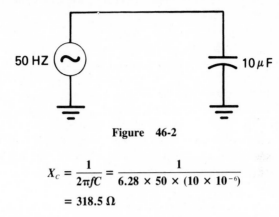

Figure   46-2

$$X_c = \frac{1}{2\pi f C} = \frac{1}{6.28 \times 50 \times (10 \times 10^{-6})}$$
$$= 318.5 \ \Omega$$

Note in this example that when the frequency of the generator was decreased, X_C increased.

EQUATION 47
CAPACITANCE
(FROM FREQUENCY AND REACTANCE)

$$C = \frac{1}{2\pi f X_C}$$

C = capacitance   F

2π = a constant

f = frequency   Hz

X_C = capacitive reactance   Ω

Equation 47 finds a useful application when we know the reactance desired but not the actual value of capacitance required to produce it. A good example of this is shown in Fig. 47-1, where we want to be certain that frequencies

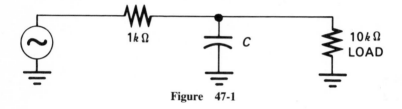

Figure 47-1

from 5 kHz and up will be shunted to ground. The lower frequencies will pass on to the 10-kΩ load. This circuit is known as a *low-pass filter*. The load shown here could represent the input to an audio amplifier. If C were to have an X_C of 100 Ω, for instance, this then would shunt most of the signal to the ground. Therefore,

$$C = \frac{1}{2\pi f X_C} = \frac{1}{6.28 \times (5 \times 10^3) \times 100} = 0.32 \ \mu F$$

A low frequency such as 50 Hz will see C as a 99.5-kΩ resistance; the current will take the path of least resistance through the 10-kΩ resistor and back to the generator.

EXAMPLE 47-1

What value of capacitance will produce a reactance of 700 Ω at 10 kHz? From Equation 47,

$$C = \frac{1}{2\pi f X_c}$$

$$= \frac{1}{6.28 \times (10 \times 10^3) \times 700}$$

$$= \frac{1}{4.4 \times 10^7}$$

$$= 0.023 \ \mu F$$

EXAMPLE 47-2

What value of capacitance will produce a reactance of 20 Ω at 10 kHz?

$$C = \frac{1}{2\pi f X_c}$$

$$= \frac{1}{6.28 \times (10 \times 10^3) \times 20}$$

$$= \frac{1}{1.3 \times 10^6}$$

$$= 0.8 \ \mu F$$

Note how much larger a capacitance is needed in order to get the X_c down to 20 Ω.

EQUATION 48
IMPEDANCE
(FROM RESISTANCE AND REACTANCE)

$$Z = \sqrt{R^2 + X^2}$$

Z = impedance Ω
R = resistance Ω
X = reactance* Ω

In a simple resistive circuit such as the circuit in Fig. 48-1, the total resistance alone makes up the impedance. Impedance is the total load that the generator sees. If the circuit contains reactive elements such as inductors

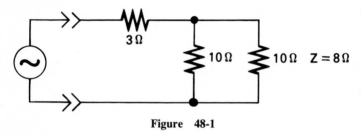

Figure 48-1

or capacitors, however, the currents and voltages are not in phase and a trigonometric approach must be used to calculate Z. Equation 48 is a direct adaptation of the Pythagorean theorem, the trigonometric calculation we used. The impedance will be the hypotenuse of a triangle in which the resistance is one leg and the reactance the other, as shown in Fig. 48-2.

$$Z = \sqrt{R^2 + X^2} = \sqrt{10^2 + 20^2} = \sqrt{500} = 22.4 \ \Omega$$

In other words, the generator in Fig. 48-2 sees a 22.4-Ω load, not 30 Ω, which it would appear to be.

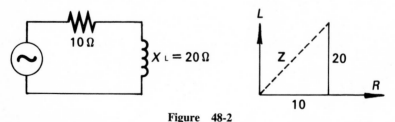

Figure 48-2

---

*Either inductive *or* capacitive.

EXAMPLE 48-1

What is the impedance of the circuit of Fig. 48-3?

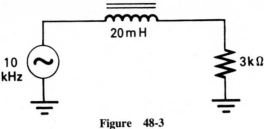

Figure   48-3

First, find the ohmic reactance of the inductor:

$$X_L = 2\pi fL = 6.28 \times (10 \times 10^3) \times 0.02 = 1256 \ \Omega$$

Then, draw the ohmic triangle (Fig. 48-4). This allows us to visualize the problem. Now solve using Equation 48:

$$Z = \sqrt{R^2 + X^2} = \sqrt{3000^2 + 1256^2} = \sqrt{1.06 \times 10^1} = 3252 \ \Omega$$

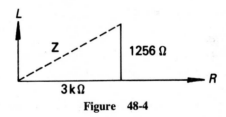

Figure   48-4

EXAMPLE 48-2

What is the impedance of the circuit in Fig. 48-5?

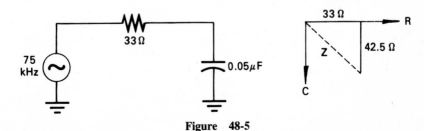

Figure   48-5

First, find the ohmic reactance of the capacitor:

$$X_c = \frac{1}{2\pi f C} = \frac{1}{6.28 \times (75 \times 10^3) \times (0.05 \times 10^{-6})} = 42.5 \ \Omega$$

Draw the ohmic triangle to visualize the problem and solve for the impedance using Equation 48:

$$Z = \sqrt{R^2 + X^2} = \sqrt{33^2 + 42.5^2} = \sqrt{1089 + 1806.3} = 53.8 \ \Omega$$

NOTE that the capacitive reactance is drawn downward or 180° from the inductive reactance.

EQUATION 49
IMPEDANCE (FROM VOLTAGE AND CURRENT)

$$Z = \frac{V}{I_T}$$

Z = impedance Ω
V = applied voltage   V
I_T = total current   A

Close inspection of Equation 49 reveals that it is very similar to Ohm's law for finding resistance. The difference is that the impedance derived is the effective AC resistance of the entire circuit, including both resistance and reactance. This formula is particularly useful in parallel AC circuits in which all of the branch currents are added up to provide I_T, from which total impedance may be calculated. This progression is depicted in Fig. 49-1.

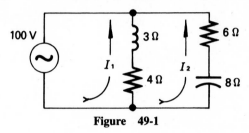

Figure   49-1

For branch 1:

$$Z_1 = 4 + j3$$
$$= 5\,\underline{/36.9°}$$
$$I_1 = \frac{V}{Z_1} = \frac{100\,\underline{/0°}}{5\,\underline{/36.9°}}$$
$$= 20\,\text{A}\,\underline{/-36.9°}$$
$$= 16 - j12$$

For branch 2:

$$Z_2 = 6 - j8$$
$$= 10\,\underline{/-53°}$$
$$I_2 = \frac{100\,\underline{/0°}}{10\,\underline{/-53°}}$$
$$= 10\,\text{A}\,\underline{/53°}$$
$$= 6 + j8$$

Totaling both branches:

$$I_T = I_1 + I_2$$
$$= 22 - j4$$
$$= 22.36 \text{ A } \underline{/-10.3°}$$
$$Z = \frac{V}{I_T}$$
$$= \frac{100}{22.36} = 4.47 \text{ }\Omega$$

NOTE The use of rectangular notation for addition and polar notation for division.

EXAMPLE 49-1

What is the impedance of the loudspeaker in Fig. 49-2?

$$Z = \frac{V}{I_T} = \frac{30}{3.75} = 8 \text{ }\Omega$$

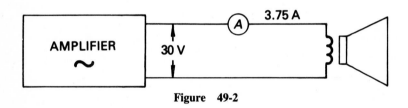

Figure 49-2

EXAMPLE 49-2

What impedance does the signal source see in Fig. 49-3?

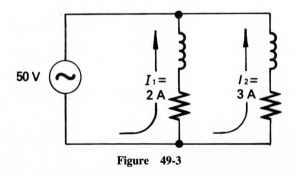

Figure 49-3

Here, assuming that the currents in both branches are in phase, they can be simply added:

$$I_\mathrm{T} = I_1 + I_2 = 2\,\mathrm{A} + 3\,\mathrm{A} = 5\,\mathrm{A}$$

Then, solving for Z:

$$Z = \frac{V}{I_\mathrm{T}}$$
$$= \frac{50}{5}$$
$$= 10\,\Omega$$

EQUATION 50 TRUE POWER		
$P_T = I^2 \times R$	P_T = true power	W
	I = current	A
	R = resistance	Ω

True power will dissipate only in pure resistance and not in reactive elements. For example, when an AC signal is applied to a resistor and an inductor, it is only in the resistor and the resistive portion of the inductor that heat is produced and power (watts) consumed. If an AC signal is applied to a *pure* inductance alone (something that doesn't really exist), it *appears* that power is dissipated. Current flows and there is a voltage present, but there is no work being done. The inductor does not even become warm. We can therefore assume that true power is not developed in any purely reactive element.

Calculating the true power of the circuit in Fig. 50-1, if we assume L to have very little DC resistance, the R entered into the equation will be only the 1000 Ω of the resistor itself. Thus we are calculating only the true power dissipation occurring in the resistance itself. The current, however, is derived by dividing the 100 V by the *impedance* of the circuit.

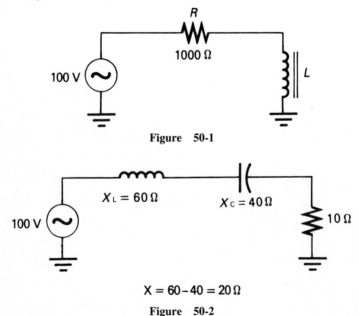

Figure   50-1

$$X = 60 - 40 = 20\,\Omega$$

Figure   50-2

EXAMPLE 50-1

Calculate the true power dissipated in Fig. 50-2.

First find the circuit impedance. Note how the capacitive and inductive reactances are subtracted from each other to arrive at X:

$$Z = \sqrt{R^2 + X^2} = \sqrt{10^2 + 20^2} = 22.4\ \Omega$$

From this, determine the circuit current:

$$I = \frac{V_A}{Z} = \frac{100}{22.4} = 4.46\ \text{A}$$

Finally, solve for the true power:

$$P_T = I^2R$$
$$= 4.46^2 \times 10 = 19.9 \times 10$$
$$= 199\ \text{W}$$

EXAMPLE 50-2

Calculate the amount of true power dissipated in Fig. 50-3.

Figure   50-3

First, find the circuit current:

$$I = \frac{V_A}{Z} = \frac{100}{10} = 10\ \text{A}$$

Then solve for true power:

$$P_T = I^2R$$
$$= 10^2 \times 10 = 100 \times 10$$
$$= 1000\ \text{W} = 1\ \text{kW}$$

In this example, note that since there were no reactive components, all of the power is true.

EQUATION 51
APPARENT POWER

$$P_A = V \times I_T$$

P_A = apparent power VA
V = applied voltage V
I_T = total current A

In a circuit which contains only pure reactance, a voltage will be present and a current flow can be measured. This makes it appear that power is being dissipated. Since the reactive element feeds voltage back to the source, however, the power is only apparent. Most actual circuits contain a combination of resistances and reactances, and a combination of both true and apparent power is produced. By analyzing Fig. 51-1 we can see this combination.

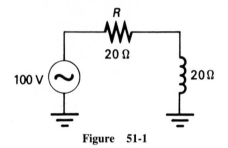

Figure 51-1

$$Z = \sqrt{R^2 + X_L^2} = \sqrt{400 + 400} = 28.3 \ \Omega$$

$$I_T = \frac{V}{Z} = \frac{100}{28.3} = 3.5 \ \text{A}$$

Apparent power	True power
$P_A = V \times I_T$	$P_T = I^2 \times R$
$= 100 \times 3.5$	$= 3.5^2 \times 20$
$= 350 \ \text{VA}$	$= 245 \ \text{W}$

Although the circuit appears to dissipate 350 W, only 245 W is actually consumed.

EXAMPLE 51-1

Calculate the apparent power of the circuit in Fig. 51-2.

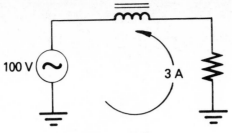

Figure 51-2

$$P_A = V \times I_T = 100 \times 3 = 300 \text{ VA}$$

EXAMPLE 51-2

Calculate the apparent power of the circuit in Fig. 51-3.

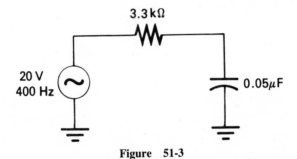

Figure 51-3

First, calculate X_C:

$$X_C = \frac{1}{2\pi fC} = \frac{1}{6.28 \times 400 \times (0.05 \times 10^{-6})} = 7962 \ \Omega$$

Second, calculate Z:

$$Z = \sqrt{R^2 + X_C^2} = \sqrt{3300^2 + 7962^2} = 8619 \ \Omega$$

Third, calculate I_T:

$$I_T = \frac{V}{Z} = \frac{20}{8619} = 2.3 \text{ mA}$$

Now, determine the apparent power:

$$P_A = V \times I_T = 20 \times (2.3 \times 10^{-3}) = 0.046 \text{ VA}$$

	EQUATION 52 POWER FACTOR
$$PF = \dfrac{P_T}{P_A}$$	PF = power factor P_T = true power W P_A = apparent power VA

By comparing the apparent power to the true power, we derive the power factor. This numerical value, which will be between 0 and 1, relates the amount of circuit reactance to circuit resistance. As an example, an incandescent lamp, which is about 99 percent resistive, will have a power factor of almost 1 and will dissipate a high wattage. On the other hand, motors, which contain reactance, will have a lower power factor and therefore will develop apparent as well as true power.

Knowing the power factor of a device such as a motor allows us to calculate the true power. For example, if $PF = 0.85$ and $P_A = 300$ VA, then:

$$P_T = PF \times P_A = 0.85 \times 300 = 255 \text{ W}$$

An alternate method for deriving the power factor is to find the cosine of the circuit phase angle theta (θ). This also provides an excellent check for Equation 52.

EXAMPLE 52-1

Will the components of the circuit in Fig. 52-1 produce a high degree of heat?

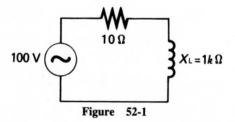

100 V   10 Ω $X_L = 1k\,\Omega$

Figure   52-1

First find the circuit current. Since the reactance is so much larger than the resistance in this circuit, the resistance may be ignored in determining the impedance:

$$I = \frac{V}{Z} - \frac{100}{1000} = 0.1 \text{ A}$$

Then calculate P_A and P_T:

$$P_A = V \times I = 100 \times 0.1 = 10 \text{ VA}$$

$$P_T = I^2R = 0.1^2 \times 10 = 0.1 \text{ W}$$

Finally, determine the power factor:

$$PF = \frac{P_T}{P_A} = \frac{0.1}{10} = 0.01$$

With a PF this low, very little true power is present, and therefore little heat is produced.

EXAMPLE 52-2

What is the power factor for the circuit in Fig. 52-2? Use the cosine of theta (θ) to check.

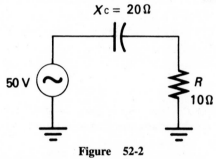

$X_C = 20\,\Omega$

50 V

R
$10\,\Omega$

Figure   52-2

First find the impedance and current of the circuit:

$$Z = \sqrt{R^2 + X_C^2} = 22.4\,\Omega$$

$$I_T = \frac{V}{Z} = 2.23 \text{ mA}$$

Then find the apparent power and the true power:

$$P_A = V \times I = 112 \text{ VA}$$

$$P_T = I^2R = 50 \text{ W}$$

Then,

$$PF = \frac{P_T}{P_A} = \frac{50}{112} = 0.45$$

To check,

$$\theta = 63.4°$$

$$\cos \theta = 0.45$$

EQUATION 53
CUTOFF FREQUENCY (INDUCTION)

$$f_{CO} = \frac{R}{2\pi L}$$

f_{CO} = cutoff frequency   Hz
R = resistance Ω
2π = a constant
L = inductance   H

When a number of different frequencies are passed through a purely resistive network as in Fig. 53-1a, although they will be attenuated, they will all emerge in the same proportion. If, however, there is an inductance in the circuit connected to ground, as in Fig. 53-1b, its reactance will selectively allow the higher frequencies to pass and the lower ones to be shunted to ground. Notice here that the resistor and the inductor form a voltage divider network with R fixed and X_L variable. This configuration is called a *high-pass filter*. The reason the circuit of Fig. 53-1b is a high-pass filter is that the lower frequencies see the inductor as a low-resistance path to ground while the higher frequencies see it as a high resistance and therefore pass on to the output.

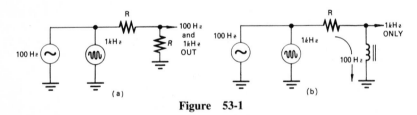

Figure 53-1

Equation 53 permits the calculation of the *lower cutoff frequency*. This is the frequency at which the signal falls to 70.7 percent of its initial amplitude; at lower frequencies in this circuit the attenuation will be even greater.

EXAMPLE 53-1

Calculate the cutoff frequency of the circuit of Fig. 53-2a.

$$f_{CO} = \frac{R}{2\pi L} = \frac{2200}{6.28 \times (33 \times 10^{-3})} = 10.616\,\text{kHz}$$

Frequencies below the f_{CO} of 10.6 kHz are reduced by more than 70.7 percent, as can be seen in Fig. 53-2b.

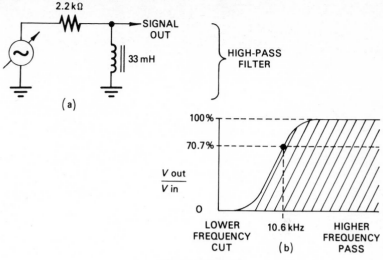

Figure 53-2

EXAMPLE 53-2

Using a 10-kΩ resistor and an inductor, design a high-pass filter for a cutoff frequency of 30 kHz. (See Fig. 53-3.)

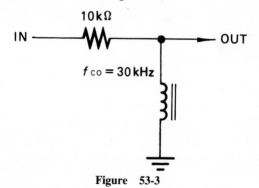

Figure 53-3

From Equation 53,

$$f_{co} = \frac{R}{2\pi L}$$

Then
$$L = \frac{R}{2\pi f_{co}} = \frac{10 \times 10^3}{6.28 \times (30 \times 10^3)} = 53 \text{ mH}$$

NOTE Reversing the position of the resistor and the coil reverses the filter action and produces a *low-pass* filter; f_{co} does not change, however.

<table>
<tr><td colspan="2">

EQUATION 54
CUTOFF FREQUENCY (CAPACITIVE)
</td></tr>
</table>

$$f_{CO} = \frac{1}{2\pi RC}$$

f_{CO} = cutoff frequency Hz
R = resistance Ω
2π = a constant
C = capacitance F

When signals of a variety of different frequencies are passed through a resistive voltage divider, as shown in Fig. 54-1a, they will be attenuated. The various frequencies, however, will still be in the same proportion. If a capacitor is connected in place of one of the resistances as shown in Fig. 54-1b, the variable effect of X_C comes into play.

The circuit in Fig. 54-1b is referred to as a *low-pass filter*. As the signal *frequency* is increased at the input, X_C begins to decrease in ohmic value until it eventually shunts the signals to ground. At a point called the *upper cutoff frequency*, which can be calculated with Equation 54, the output signal will be 70.7 percent in amplitude of the input signal. At higher frequencies it will be even less.

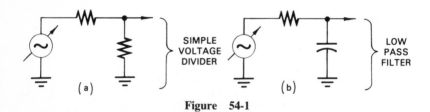

Figure   54-1

EXAMPLE 54-1

Calculate the cutoff frequency of the circuit in Fig. 54-2a.

$$f_{CO} = \frac{1}{2\pi RC} = \frac{1}{6.28 \times (10 \times 10^{3}) \times (220 \times 10^{-12})} = 72.380\,\text{kHz}$$

EXAMPLE 54-2

Reversing the position of the components in Fig. 54-2 results in a high-pass filter; f_{CO} remains the same. Find f_{CO} for the circuit of Fig. 54-3a.

$$f_{CO} = \frac{1}{2\pi RC} = \frac{1}{6.28 \times (10 \times 10^{3}) \times (220 \times 10^{-12})} = 72.380\,\text{kHz}$$

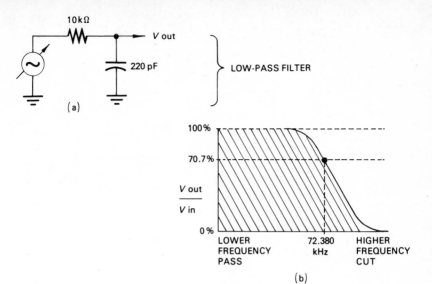

Figure 54-2

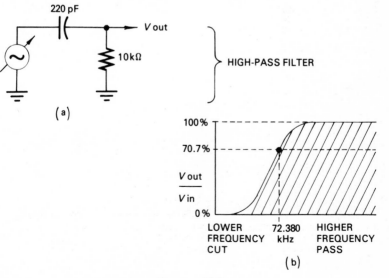

Figure 54-3

EQUATION 55
RESONANT FREQUENCY

$$f_0 = \frac{1}{2\pi\sqrt{LC}}$$

f_0 = resonant frequency   Hz
2π = a constant
L = inductance   H
C = capacitance   F

Knowing that as frequency increases X_L increases and X_C decreases (and vice versa), we can conclude that at some specific frequency they will be equal. Let us first examine this behavior in a circuit in which the inductor and the capacitor are in series. At the resonant frequency, X_L equals X_C and thus the two cancel and produce a total reactance of 0. This results in maximum current flow, since the impedance is only the resistance and $I = V/R$. On the other hand, in a circuit where the inductor and capacitor are in parallel, when the current through the capacitor cancels the current through the inductor, there will be little or no current flow. Here the impedance must be high, because $Z = V/I$ and I is low.

In summary, then, in a series resonant circuit, current is high and is limited only by any pure resistance present. In a parallel resonant circuit, current is at a minimum and passes mainly through any remaining pure resistance in parallel with the circuit. In either case, if the driving signal is shifted off resonance, the circuits begin to appear either capacitive or inductive and resonance does not exist. In addition to Ohm's law, Equation 55 is probably the most significant in the field of electronics. It would be wise for you to gain a total understanding of its function.

EXAMPLE 55-1

Find the resonant frequency and current flow in Fig. 55-1.

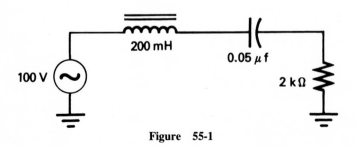

Figure 55-1

First, find f_0:

$$f_0 = \frac{1}{2\pi\sqrt{LC}} = \frac{1}{6.28\sqrt{(200 \times 10^{-3}) \times (0.05 \times 10^{-6})}} = 1592\,\text{Hz}$$

Then, because the reactances cancel at the resonant frequency:

$$Z = 2\text{k}\Omega \text{ at } 1592\,\text{Hz}$$

Thus,

$$I = \frac{V}{Z} = \frac{100}{2 \times 10^3} = 50\,\text{mA}$$

EXAMPLE 55-2

Find the resonant frequency and current flow of the circuit in Fig. 55-2.

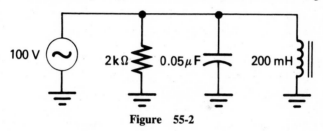

Figure 55-2

The component values here are the same as those from Example 55-1. Thus,

$$f_0 = 1592\,\text{Hz} \qquad \text{and} \qquad I_T = 50\,\text{mA}$$

Because the reactances cancel at f_0, the circuit is equivalent to the circuit of Fig. 55-1, and the current is a result of only the 2-kΩ resistor. Note that this current is through the resistor only. If the resistor were not in the circuit, there would essentially be zero current at f_0.

EQUATION 56
BANDWIDTH

$$BW = \frac{f_0}{Q}$$

BW = bandwidth   Hz
f_0 = resonant frequency   Hz
Q = quality factor

Bandwidth refers to all frequencies between a given lower and a given upper cutoff frequency. The circuit shown in Fig. 56-1 is resonant at 1000 Hz and has a bandwidth of 20 Hz. When a circuit is operated at its resonant frequency, the bandwidth will be evenly distributed on either side, in this case, 10 Hz below and 10 Hz above. A result of this will be that all signals from 990 Hz to 1010 Hz will arrive at the load at 70.7 percent at least of their initial amplitude; all others will be less. The factor that determines this width is the Q of the tuned circuit. Q here refers to the quality, or lack of DC resistance, in the coil-capacitor combination. Q factors of less than 10 are said to be low. Note that as the Q factor increases, the bandwidth decreases.

Rearranging Equation 56: $Q = f_0/BW$; therefore, in Fig. 56-1, Q = 50.

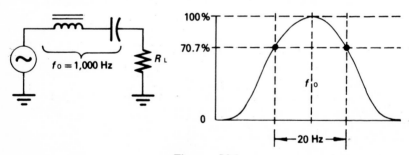

Figure 56-1

EXAMPLE 56-1

At what frequency will the voltage be at maximum across the load in Fig. 56-2? The voltage across R_L will be maximum at series resonance. Thus,

$$f_0 = \frac{1}{2\pi\sqrt{LC}} = \frac{1}{6.28\sqrt{(40 \times 10^{-3}) \times (0.05 \times 10^{-6})}} = 3561\ Hz$$

110   AC CIRCUITS

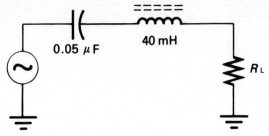

Figure 56-2

EXAMPLE 56-2

If the Q of the tuned circuit in Example 56-1 is 54, plot the resulting pass band curve.

First determine the bandwidth:

$$BW = \frac{f_0}{Q} = \frac{3561}{54} = 66 \text{ Hz}$$

The resulting curve will pass through the 70.7 percent points at 33 Hz above and below f_0, as shown in Fig. 56-3. Care must be taken not to allow the load to become too high in resistance, as this will reduce the Q of the circuit and broaden the bandwidth.

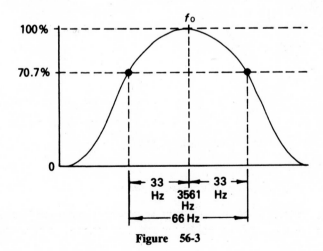

Figure 56-3

EQUATION 57
IMPEDANCE MATCHING

$$a = \sqrt{\frac{Z_P}{Z_L}}$$

a = turns ratio

Z_P = primary impedance $\quad \Omega$

Z_L = load impedance $\quad \Omega$

Efficient transfer of power from one stage to another is accomplished only when both stage's impedances are matched. An example of this is the loss in volume when a low-impedance microphone is fed into a high-input-impedance amplifier. Although the sound quality will be acceptable, the volume level will be too low. To compensate for this mismatch, a step-up transformer may be inserted to correct the situation. In this example, the microphone impedance is Z_P and it is working into the input load Z_L. Taking the square root of the quotient of Z_P and Z_L produces the turns ratio required. If a transformer with this ratio is incorporated into the circuit, the amplifier's high input impedance will be reflected back to the microphone and will appear to be low. The microphone thus appears to be working into its required low-impedance load. Equation 57 is also applicable to output circuits, where Z_P might represent the collector portion of a transistor circuit and Z_L the load being driven. An example of such a load might be a loudspeaker, relay, or other such device.

EXAMPLE 57-1

Determine the required transformer turns ratio to match the microphone to the amplifier's input in Fig. 57-1.

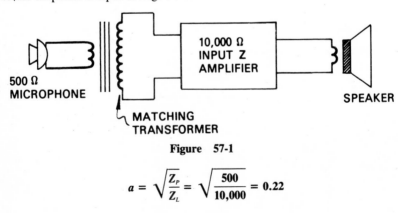

500 Ω
MICROPHONE

10,000 Ω
INPUT Z
AMPLIFIER

SPEAKER

MATCHING
TRANSFORMER

Figure 57-1

$$a = \sqrt{\frac{Z_P}{Z_L}} = \sqrt{\frac{500}{10,000}} = 0.22$$

The necessary turns ratio is 0.22:1 or 1:4.5.

EXAMPLE 57-2

Match the output of the vacuum tube to the loudspeaker in Fig. 57-2.

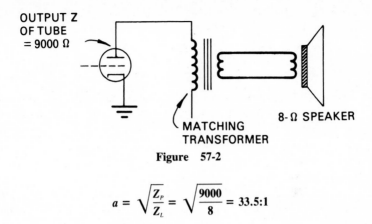

OUTPUT Z
OF TUBE
= 9000 Ω

8-Ω SPEAKER

MATCHING
TRANSFORMER

Figure 57-2

$$a = \sqrt{\frac{Z_P}{Z_L}} = \sqrt{\frac{9000}{8}} = 33.5:1$$

Thus, the primary should have 33.5 times the number of turns the secondary has.

ACTIVE DEVICES

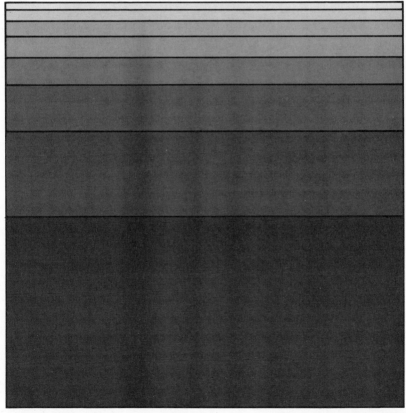

EQUATION 58		
DIODE FUNCTION		
$V_S = V_L + V_F$	V_S = source voltage	V
	V_L = load voltage	V
	V_F = diode forward voltage	V

A diode performs its function as a result of the electrical action taking place at its internal junction. The junction itself acts as a one-way switch. By this we mean that current may pass through in one direction only. Unlike the operation of a standard switch, however, the turn-on action is not immediate. Very low applied voltages do not activate the junction and current does not flow. Increasing the applied voltage will increasingly turn the junction on. If the diode is a silicon type, when 0.7 V (V_F) is reached across the diode, it is considered to be fully on. Any further increase in V_S will not increase V_F appreciably beyond 0.7 V. (See Fig. 58-1.)

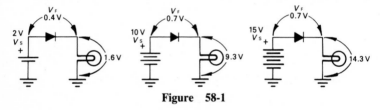

Figure   58-1

From Equation 58 we see that the load voltage plus the diode drop will be equal to the source voltage. Once the 0.7 V is reached, any increase in V_S will simply develop across the load.

EXAMPLE 58-1

In the circuit of Fig. 58-2 is the diode fully turned on? What is the battery voltage?

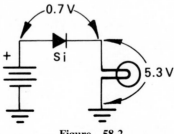

Figure   58-2

The diode is fully on, as $V_F = 0.7$ V. The battery voltage can be found by using Equation 58:

$$V_S = V_L + V_F = 5.3 + 0.7 = 6 \text{ V}$$

EXAMPLE 58-2

What is the current flowing through the load in Fig. 58-3?

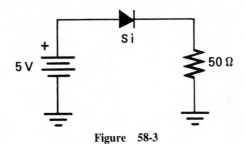

Figure 58-3

From Equation 58, V_L must be 4.3 V (assume the diode to be fully on). Current through the resistor equals the total circuit current. Thus,

$$I = \frac{V}{R} = \frac{4.3}{50} = 86 \text{ mA}$$

EQUATION 59
ZENER RESISTANCE (HIGH)

$$R_{ZH} = \frac{V_s - V_z}{I_L + I_{ZK}}$$

R_{ZH} = series resistor for Zener diode	Ω
V_s = source voltage	V
V_z = Zener voltage	V
I_{ZK} = Zener current	A
I_L = load current	A

The primary purpose of a Zener diode is to provide a reasonably regulated voltage which another circuit may require for its operation. Without the Zener, the voltage could vary as a result of one of two factors: first, a varying source voltage; second, a varying load current. Equation 59 is specifically geared to accommodate the second, as this usually poses the greater problem; I_L is given for the maximum load to be driven. A typical Zener regulator is shown in Fig. 59-1.

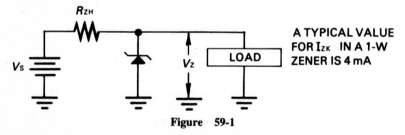

Figure 59-1

A TYPICAL VALUE FOR I_{ZK} IN A 1-W ZENER IS 4 mA

Equation 59 will give us the highest possible value for R_{ZH} that will both keep the Zener just on and also supply the required load current. If the load slightly exceeds I_L, however, Zener regulation will cease, and an improper lower voltage will result at the load. Nevertheless, this approach will allow the coolest operation of the Zener and at the same time require the lightest standby load on the battery.

EXAMPLE 59-1

Find the value of R_{ZH} in Fig. 59-2 if the Zener is a 15-V unit with a required maintenance current I_{ZK} of 4 mA.*

$$R_{ZH} = \frac{V_s - V_z}{I_L + I_{ZK}} = \frac{25 - 15}{0.06 + 0.004} = 156 \ \Omega$$

*This value is typical for most 1-W units.

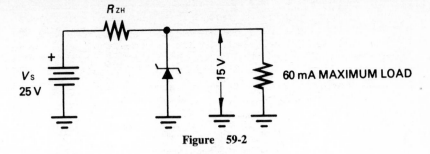

Figure   59-2

EXAMPLE 59-2

Since any Zener calculation should always be made for the maximum load to be driven, find R_{ZH} in Fig. 59-3 is $V_Z = 6$ V.

$$R_{ZH} = \frac{V_s - V_Z}{I_L + I_{ZK}} = \frac{12 - 6}{0.05 + 0.004} = 111 \ \Omega$$

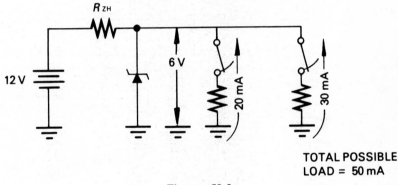

TOTAL POSSIBLE
LOAD = 50 mA

Figure   59-3

	EQUATION **60**
	ZENER RESISTANCE (LOW)

$$R_{ZL} = \frac{V_S - V_Z}{0.5 \times I_{ZM}}$$

R_{ZL} = series resistor for Zener diode   Ω
V_S = source voltage   V
V_Z = Zener voltage   V
I_{ZM} = maximum Zener current   A

Zener diodes, often called *reference diodes,* are available in a wide range of voltage ratings, from 0.4 to 200 V. The two most commonly used power ratings are ½ W and 1 W. Equation 60 will yield the lowest value of series resistance that will drive the Zener to one-half its power rating. We will consider this a maximum condition. Running a Zener diode at close to its full limit could result in a premature malfunction. Using this form of calculation does not require that we incorporate any load demands. This is because R_{ZL} is calculated at its lowest value and could not safely pass on any more current even if the load demanded it.

Comparing Equation 59 to this equation leads us to this conclusion: Where the lightest possible load on the source is required and the load is fairly constant, use Equation 59. If V_S is substantial and the load is quite erratic, especially in the high-current region, use Equation 60.

EXAMPLE 60-1

A 1-W Zener diode with a Zener voltage of 9 V is used in the circuit of Fig. 60-1. Calculate the lowest value that R_{ZL} should be.

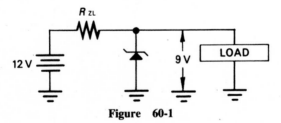

Figure 60-1

First determine I_{ZM}:

$$I = \frac{P}{V} = \frac{1\,W}{9} = 111\ mA$$

Now solve for R_{ZL}:

$$R_{ZL} = \frac{V_S - V_Z}{0.5 \times I_{ZM}} = \frac{12 - 9}{0.5 \times (111 \times 10^{-3})} = 54\ \Omega$$

NOTE The current limit of 0.555 A set by R_{ZL} and the 3 V dropped across it demands that the load not draw more than 55 mA less 4 mA, or 51 mA.

EXAMPLE 60-2

A 500-mW Zener diode with a Zener voltage of 6 V is used in the circuit of Fig. 60-2. Find R_{ZL}.

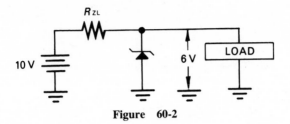

Figure 60-2

First find I_{ZM}:

$$I_{ZM} = \frac{P}{V_z} = \frac{0.5}{6} = 83.3 \, \text{mA}$$

Now solve for R_{ZL}:

$$R_{ZL} = \frac{V_s - V_z}{0.5 \times I_{ZM}} = \frac{10 - 6}{0.5 \times (83.3 \times 10^{-3})} = 96 \, \Omega$$

EQUATION 61
TRANSISTOR CURRENTS

$$I_E = I_B + I_C$$

I_E = emitter current   A*

I_B = base current   A*

I_C = collector current   A*

The relationships between the various currents in a BJT (bipolar junction transistor) give us an insight into its functioning. Equation 61 states that the largest current present is in the emitter and it comprises both the collector current and the base current. This current distribution is depicted in Fig. 61-1.

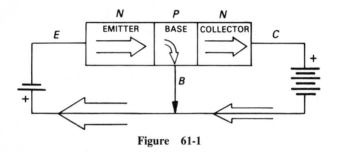

Figure 61-1

Note that if the base lead is disconnected, the reverse-biased collector will stop all current flow. We can also see that the collector current will never be as great as the emitter current. The relationship between I_E and I_C is called alpha (α). Alpha is equal to I_C/I_E and is never greater than 1. An actual common-base circuit (with base grounded) would contain resistances to limit current flows to a safe level, as shown in Fig. 61-2.

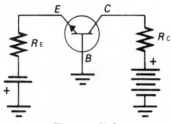

Figure 61-2

*Typically, these currents are given in smaller units, such as milliamperes. The currents do not have to be in basic units so long as similar units are used for quantities to be added.

EXAMPLE 61-1

If the emitter current is 10.3 mA and the collector current is 10 mA, what is the current flowing from base to ground in Fig. 61-3?

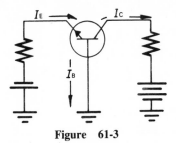

Figure 61-3

By rearranging the variables in Equation 61, we can solve for I_B:

$$I_B = I_E - I_C = 10.3 - 10 = 300 \ \mu A$$

EXAMPLE 61-2

Find the collector current in Fig. 61-4 if $I_B = 0.2$ mA. Assume the base emitter junction acts as a silicon diode.

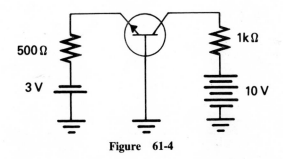

Figure 61-4

First determine I_E:

$$I_E = \frac{V}{R} = \frac{3 \ V - 0.7 \ V}{500 \ \Omega} = 4.6 \ mA$$

Solving Equation 61 for I_C:

$$I_C = I_E - I_B = 4.6 - 0.2 = 4.4 \ mA$$

EQUATION 62
TRANSISTOR GAIN

$$I_C = \beta \times I_B$$

I_C = collector current   mA*

β = current gain

I_B = base current μA*

The common-emitter (grounded-emitter) configuration shown in Fig. 62-1 is the most widely used of the three possible bipolar transistor current types. Connecting the transistor in this manner provides us with both voltage and current gain at the output (in the collector circuit).

The beta factor tells us what the current gain of the transistor is. Low-gain transistors have a β in the range of 10 to 50. Others go as high as 1000. If, for instance, in Fig. 62-1 the transistor has a beta of 100 and the signal at the base causes 2 mA to flow, the resulting output current flow in the collector will be 100 × 2 mA, or 200 mA. If we now substitute a higher-gain transistor with a beta of 200, the collector current will be 400 mA with the same input signal. All this, of course, is possible only if V_{CC} and R_C do not limit the higher output currents.

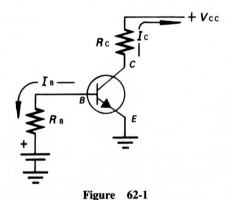

Figure   62-1

EXAMPLE 62-1

The lamp in the circuit of Fig. 62-2 requires 200 mA to be fully lit. Will it be lit if the transistor has an input signal of 1.5 V?

*Typical units.

First find the base current. In finding I_B remember the base-emitter junction drop.

$$I_B = \frac{V_{BB} - V_{BE}}{R_B} = \frac{1.5 - 0.7}{2 \times 10^3} = 400 \ \mu A$$

Now solve for I_C:

$$I_C = \beta \times I_B = 500 \times (400 \times 10^{-6}) = 200 \ mA$$

200 mA will flow in the collector portion of the circuit, and the lamp will be fully lit.

EXAMPLE 62-2

What input current will be required to produce a collector current of 17.25 mA in the circuit?

By rearranging the variables in Equation 62 we can solve for I_B:

$$I_B = \frac{I_C}{\beta} = \frac{17.25 \times 10^{-3}}{75} = 230 \ \mu A$$

NOTE At this point we could use R_B and 230 μA to calculate the value of V_{BB}. Here we would have to *add* the 0.7 V back into the equation.

EQUATION 63
TRANSISTOR LOAD LINE

$V_{CC} = I_C \times R_C + V_{CE}$	V_{CC} = collector supply voltage V I_C = collector current A R_C = collector resistor Ω V_{CE} = voltage between collector and emitter V

The load line equation shows the relationship of the functioning transistor to the circuit within which it is operating (see Fig. 63-1).

Another way of stating Equation 63 is $V_{CC} = V_C + V_{CE}$, since $V_C = I_C \times R_C$. Two extreme cases will illustrate this point. First, when the transistor is completely cut off (no input signal), $I_C \times R_C$ is 0 and all the voltage is across the open transistor: thus 10 V = 0 V + 10 V. Second, when the transistor is fully on (saturated) because of a large input signal, $I_C \times R_C$ is 10 V and there is no voltage across the closed transistor: thus, 10 V = 10 V + 0 V. In a final, intermediate case with the transistor 50 percent on, which is referred to as class A operation, 10 V = 5 V + 5 V. Any operation between cutoff and saturation is said to be in the *active region*.

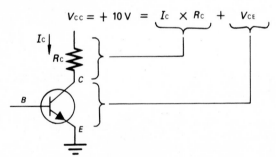

Figure 63-1

EXAMPLE 63-1

What is V_{CC} in Fig. 63-2 if the current measured through R_C is 1.2 mA and the voltage measured from collector to emitter is 7.4 V?

$$V_{CC} = I_C \times R_C + V_{CE}$$
$$= (1.2 \times 10^{-3}) \times (3 \times 10^3) + 7.4$$
$$= 3.6 + 7.4$$
$$= 11 \text{ V}$$

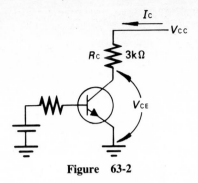

Figure 63-2

EXAMPLE 63-2

What must the voltage (V_{CE}) at the output terminal be in Fig. 63-3?

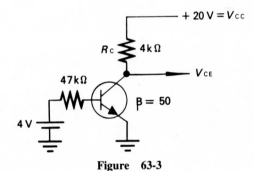

Figure 63-3

First determine the base current, taking into account the 0.7-V drop across the base-emitter junction:

$$I_B = \frac{V_{BB}}{R_B} = \frac{4 - 0.7}{47 \times 10^3} = 70.2 \ \mu A$$

Then find I_C:

$$I_C = \beta \times I_B = 50 \times (70.2 \times 10^{-6}) = 3.5 \ mA$$

If $V_{CC} = I_C \times R_C + V_{EC}$, then solving for V_{CE} gives:

$$\begin{aligned}
V_{CE} &= V_{CC} - I_C \times R_C \\
&= 20 - (3.5 \times 10^{-3}) \times (4 \times 10^3) \\
&= 20 - 14 \\
&= 6 \ V
\end{aligned}$$

EQUATION 64
UJT FUNCTION

$V_P = \eta V_{BB} + V_{DD}$	V_P = peak turnon voltage	V
	η (eta) = internal resistance ratio	
	V_{BB} = supply voltage	V
	V_{DD} = internal diode voltage drop	V*

Before gaining a full understanding of this equation we must first analyze the term η (eta) and the UJT (unijunction transistor) itself. Unlike a conventional BJT, the UJT cannot be turned on gradually but has more of a "snap" switch action.

Fig. 64-1a depicts point X as the center of voltage divider R_1-R_2. The ratio of R_2 to $R_1 + R_2$ is eta. Thus, if eta is 0.65 and V_{BB} is 10 V, it will take an input of 10 V × 0.65, or 6.5 V, at point X to switch the transistor on. We must now, according to Equation 63, add another 0.6 V to overcome the diode voltage drop when applying the input trigger signal to the emitter. V_P at the emitter thus equals the diode voltage drop plus the voltage at point X. Any value above V_P at the emitter will turn the transistor on. The actual current discharge occurs from the emitter through R_{B1} to ground, as shown in Fig. 64-1b. Once the UJT is turned on, V_P must fall to about one-sixth its initial value before the transistor will turn off.

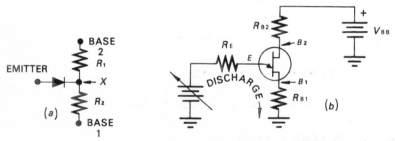

Figure   64-1

EXAMPLE 64-1

Determine the value of V_P required to turn on the UJT in Fig. 64-2. Note: Eta ranges from 0.5 to 0.8 for different units. Its specific value may be found in most data manuals.

$$V_P = \eta V_{BB} + V_{DD} = 0.74 \times 20 + 0.6 = 15.4 \text{ V}$$

*The typical value is 0.6 V.

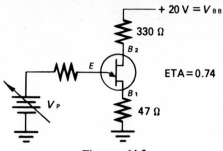

Figure   64-2

EXAMPLE 64-2

When the UJT in the circuit of Fig. 64-3 is turned on, the voltage drop from E to B_1 is approximately 2 V. Calculate V_P required for turnon and the voltage across R_E once the UJT is turned on.

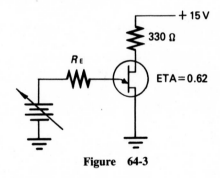

Figure   64-3

First find the turnon voltage from Equation 64:

$$V_P = \eta V_{BB} + V_{DD} = 0.62 \times 15 + 0.6 = 9.9 \text{ V}$$

Now find the voltage across R_E:

$$V_{RE} = V_P - 2 \text{ V} = 7.9 \text{ V}$$

$$f = \frac{1}{RC \times \ln(V_{IN}/(V_{IN} - V_{P3}))}$$

f = frequency of oscillation	Hz
R = timing resistor	Ω
C = timing capacitor	F
V_{IN} = source voltage	V
V_P = turnon voltage	V
L_N = natural log	

By utilizing the fast switching action of the UJT and the RC time constant of a capacitor, you can construct an excellent sawtooth oscillator. Figure 65-1 shows how the action of the transistor is analogous to manually opening and closing a switch. When the voltage across the capacitor in Fig. 65-1b reaches the transistor's trigger level V_P, the emitter-to-base-1 diode conducts. This action is very rapid, and the capacitor is immediately discharged. The UJT turns off and the capacitor again begins to charge. The value of R should be no lower than 2200 Ω and no higher than 500,000 Ω for proper transistor action to be maintained. The time required for charge incorporates both the standard RC time constant and the natural logarithm function as applied to the eta value. This combination, when inverted, yields the oscillator's frequency.

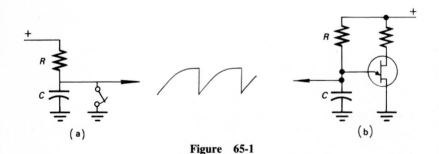

(a)

(b)

Figure 65-1

EXAMPLE 65-1

Calculate the frequency of the UJT oscillator in Fig. 65-2. Note that a spike output having the same frequency is also available at B_1.

$$V_P = \eta V_{BB} + V_{DD} = 0.7 \times 12 + 0.6 = 9 \text{ V}$$

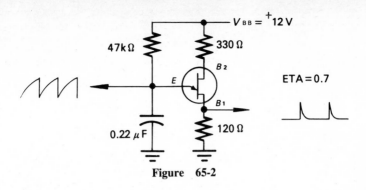

Figure 65-2

As Equation 65 requires V_P, you must first derive this from Equation 64. Now solve for f:

$$f = \frac{1}{RC \times \ln(V_{IN}/(V_{IN} - V_P))}$$

$$= \frac{1}{(47 \times 10^3) \times (0.22 \times 10^{-6}) \times \ln(12/(12 - 9))}$$

$$= \frac{1}{0.01034 \times \ln 4}$$

$$= \frac{1}{0.01034 \times 1.386}$$

$$= 71.6\,\text{Hz}$$

EQUATION 66
FET TRANSCONDUCTANCE

$$g_{fs} = \frac{\Delta I_{DP}}{\Delta V_{GS}}$$

g_{fs} = forward transconductance   S

I_{DP} = drain current   A

V_{GS} = gate-to-source voltage   V

The forward transconductance value of a field-effect transistor gives us an indication of the device's amplifying potential. Equation 66 states that higher values of g_{fs} are obtained when smaller changes in input voltage produce the same current changes in the output. Let us compare two transistors in order to examine this effect. In transistor A the input signal shifts from -1 to -2 V and, as a result, produces a current shift from 6 mA to 4 mA. By application of Equation 66 we will derive a transconductance of 0.002 S (the unit of measure is siemens), which is usually expressed as 2000 μS. Transistor B is of a higher-gain type, and changing its input from -1 to -1.4 V shifts the drain current from 6 mA to 4 mA. For this transistor we find that the transconductance is 0.005 S or 5000 μS. The values obtained relate directly to the gain of the device and, as a result, may be used in calculating overall amplifier stage voltage gain. Note: The term Δ (delta) indicates a change or a shift in value.

EXAMPLE 66-1

Calculate the forward transconductance of the field-effect transistor in Fig. 66-1.

$$g_{fs} = \frac{\Delta I_{DP}}{\Delta V_{GS}} = \frac{7\,\text{mA} - 4\,\text{mA}}{1\,\text{V} - 0.5\,\text{V}} = 6000\ \mu\text{S}$$

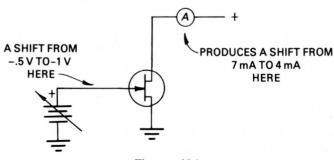

A SHIFT FROM −.5 V TO −1 V HERE

PRODUCES A SHIFT FROM 7 mA TO 4 mA HERE

Figure 66-1

EXAMPLE 66-2

Show that a transistor for which ΔI_{DP} is 4 mA and ΔV_{GS} is 0.55 V will have a higher gain than the one listed in Example 66-1.

$$g_{fs} = \frac{\Delta I_{DP}}{\Delta V_{GS}} = \frac{4 \text{ mA}}{0.55 \text{ V}} = 7273 \ \mu\text{S}$$

This is a higher value than that of the transistor in Example 66-1. Therefore, it will be of higher gain.

EQUATION 67
FET STAGE GAIN

$$A_V = g_{fs} \times R_L$$

A_V = voltage gain   V

g_{fs} = transconductance   S

R_L = load resistor Ω

Application of a changing voltage to a field-effect transistor's gate will result in a variation in its drain current. If there is a load resistor placed in the drain circuit, this changing current will produce voltage variations across it. Fig. 67-1 illustrates this action; V_{RL} is the drain resistor voltage.

You can see by looking at Equation 67 that by either increasing the value of R_L or utilizing a higher-gain FET (higher g_{fs}) you will produce a higher voltage at R_L. You will also notice in Fig. 67-1 that a small amount of negative bias voltage has been applied to the gate. The FET, unlike the BJT, with no input signal is already turned on. We must, therefore, partially turn it off with reverse gate bias. The input signal itself may now turn the device more on or off.

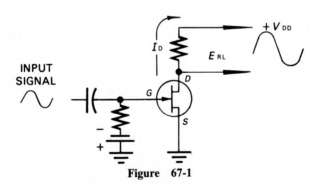

Figure 67-1

EXAMPLE 67-1

If a field-effect transistor has a g_{fs} of 2000 μS and employs a 10-kΩ load resistor, calculate the voltage gain of the stage.

$$A_v = g_{fs} \times R_L = (2000 \times 10^{-6}) \times (10 \times 10^3) = 20$$

EXAMPLE 67-2

A 300-mV signal is applied to the amplifier in Fig. 67-2. What is the amplitude of the output signal?

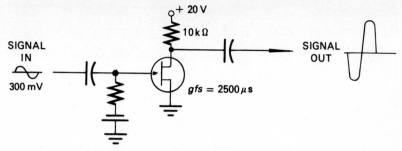

Figure 67-2

First, using the g_{fs} of the transistor, calculate the voltage gain:

$$A_V = g_{fs} \times R_L = (2500 \times 10^{-6}) \times (10 \times 10^3) = 25$$

Since the voltage gain is the ratio between V_{OUT} and V_{IN}, it will now be easy to determine the output voltage:

$$V_{OUT} = A_V \times V_{IN} = 25 \times 300\,mV = 7.5\,V$$

EQUATION 68
TUBE TRANSCONDUCTANCE

$$g_m = \frac{\Delta I_P}{\Delta V_g}$$

g_m = transconductance	S
I_P = plate current	A
V_g = grid voltage	V

The transconductance of a vacuum tube gives us an indication of the device's amplifying potential. The higher the value, the higher the amplifying gain potential. Typical values of g_m for a triode vary from 1 to 10 mS. Transconductance itself is the ratio of changes in plate current to the changes in input grid voltage that causes them. At this point we should obtain an understanding of plate resistance and a factor called *mu*. Plate resistance is an ohmic value relative to the internal plate structure of each tube type. When the internal plate resistance (r_p) is multiplied by the transconductance (g_m), the resultant factor is mu. Therefore, mu is the tube's absolute amplifying factor. It is important here to note that in analyzing tube gain we incorporate the internal plate resistance, whereas with a field-effect transistor the trans-conductance factor is used independently.

EXAMPLE 68-1

A 2-V shift at the grid of a particular vacuum tube results in a plate current shift from 10 mA to 30 mA. What is the transconductance?

$$g_m = \frac{\Delta I_p}{\Delta V_g} = \frac{30\,\text{mA} - 10\,\text{mA}}{2\,\text{V}} = \frac{20\,\text{mA}}{2\,\text{V}}$$

$$= 10\,\text{mS}$$

EXAMPLE 68-2

The tube in Example 68-1 has an internal plate resistance of 6000 Ω. Calculate the amplification factor (mu). (Note: the symbol μ represents mu.)

$$\mu = r_p \times g_m = 6000 \times (10 \times 10^{-3}) = 60$$

EQUATION 69
TUBE STAGE GAIN

$$A_V = \frac{\mu \times R_L}{r_p + R_L}$$

A_V = voltage gain
μ = tube gain
R_L = load resistor $\quad \Omega$
r_p = plate resistance $\quad \Omega$

Equation 68 describes the derivation of the term μ, which is a tube's amplification factor. A tube by itself, however, is useless as an amplifying device until it is incorporated into a circuit containing some external resistance. It is across this external resistance that the plate's changing current will produce a voltage swing, as in the simplified amplifier circuit in Fig. 69-1.

When you compare Equation 69 to the schematic, you will see that as the value of R_L is increased, the voltage swing will increase. A_V, the stage voltage gain, is easily controlled by varying the size of this resistor, which is usually within the range of 5 kΩ to 220 kΩ.

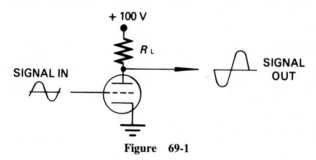

Figure   69-1

EXAMPLE 69-1

Calculate the stage gain of the amplifier in Fig. 69-2 and also determine the rms voltage of the output signal.

First determine mu by using g_m and r_p:

$$\mu = g_m \times r_p = (5 \times 10^{-3}) \times (10 \times 10^3) = 50$$

Now determine the stage gain, using Equation 69:

$$A_V = \frac{\mu \times R_L}{r_p + R_L} = \frac{50 \times (100 \times 10^3)}{(10 \times 10^3) + (100 \times 10^3)} = \frac{5 \times 10^6}{110 \times 10^3}$$

$$= -45.5$$

Finally, determine the output voltage:

$$V_{OUT} = V_{IN} \times A_V = 4 \times 45.5 = 182 \text{ V rms}$$

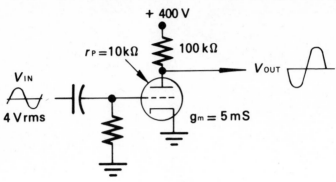

Figure   69-2

EQUATION 70
GRID CUTOFF

$$V_{CO} = \frac{-V_B}{\mu}$$

V_{CO} = grid cutoff voltage   V
V_B = plate supply voltage   V
μ = tube amplification factor

The grid of a vacuum tube is a meshlike screen that separates the cathode from the plate. Its purpose is to control the flow of electrons that stream from the cathode to the highly positive plate. This action is depicted in Fig. 70-1. With no voltage at the control grid, this flow is unhindered and the tube is fully on. To take advantage of the tube's turnon/cutoff action, it is necessary to place a small negative bias voltage on the grid. By using Equation 70 you can determine negative voltage that will just cut the flow of electrons from cathode to plate to 0; this is the *cutoff voltage* (V_{CO}). What happens is that the negative grid repels the electrons back to the cathode. Application of 0.5 times V_{CO} to the grid will allow *approximately* one-half the plate current to flow. With the tube one-half on, it is now possible to apply an additional positive or negative input signal and cause either an increase or decrease in plate current flow. Such an action is found in class A amplifiers.

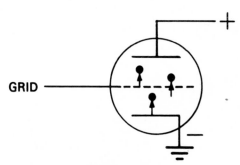

Figure 70-1

EXAMPLE 70-1

What negative grid voltage is required to make the plate current equal to zero in Fig. 70-2?

$$V_{CO} = \frac{-V_B}{\mu} = \frac{-300}{50} = -6 \text{ V}$$

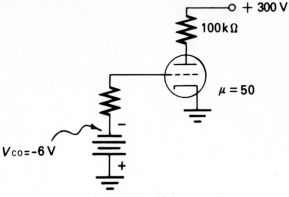

Figure 70-2

EXAMPLE 70-2

Class C radio-frequency (RF) amplifiers are usually operated with their grids at 3 times cutoff. Determine the required grid bias for class C operation of the RF amplifier in Fig. 70-3.

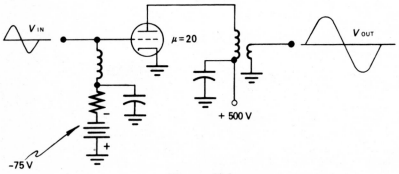

Figure 70-3

First determine V_{CO}:

$$V_{CO} = \frac{-V_B}{\mu} = \frac{-500}{20} = -25 \text{ V}$$

For class C operation, multiply V_{CO} by 3:

$$3 \times V_{CO} = -75 \text{ V}$$

CIRCUIT ANALYSIS

EQUATION 71
DIODE REVERSE BIAS

$V_{DT} = V_R + V_F$	V_{DT} = turnon voltage	V
	V_R = reverse voltage bias	V
	V_F = diode forward voltage drop	V

Diodes, whether silicon or germanium, are usually operated in one of two modes: first, as diodes by themselves, where the forward voltage that turns the junction on has to overcome only V_F (which is 0.7 V for silicon and 0.3 V for germanium); second, with a reverse bias voltage applied, in which case this reverse voltage must also be overcome before turnon occurs. Equation 71 relates to this action. It states that for the diode to turn on, the forward voltage V_{DT} must overcome both the reverse bias V_R and the diode voltage drop V_F. You might ask the question why we would want to apply this reverse voltage. One of many useful functions is shown in Fig. 71-1, where, by applying various reverse voltages, we can produce a variety of signals from one input.

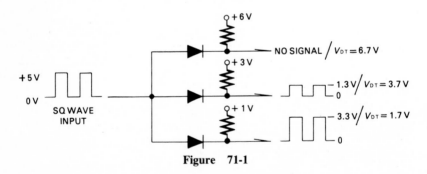

Figure   71-1

EXAMPLE 71-1

What must the amplitude of the input pulse be in Fig. 71-2 in order to present a signal at the output? Why might you need the reverse bias network?

$$V_{DT} = V_R + V_F = 3\,V + 0.7\,V = 3.7\,V$$

The high V_{DT} resulting from the reverse bias network allows only the pulse itself to pass if the line from the UJT is long and has a high level of interference voltage upon it.

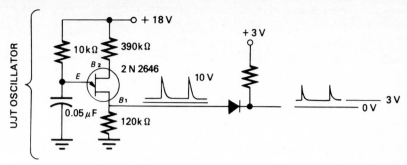

Figure 71-2

EXAMPLE 71-2

If the pulse input to the circuit in Fig. 71-2 is 10 V, what is the amplitude of the output spikes?

To find the output voltage, simply subtract the turnon voltage from the input:

$$V_{IN} - V_{DT} = V_{OUT}$$
$$10 - 3.7 = 6.3 \text{ V}$$

EQUATION 72
FET BIAS

$$V_{GS} = -V_{RS}$$

V_{GS} = gate-to-source voltage   V
V_{RS} = source resistor voltage   V

We often have to bias the gate of a field-effect transistor with some negative voltage. One method employs a separate power supply and is seldom used. An alternate method, called self-bias, simply requires the addition of a source resistance across which a voltage V_{RS} will develop. This voltage, in turn, will reflect itself to the gate as a negative voltage and hold the FET at some partial conduction point. Shown in Fig. 72-1 is a set of characteristic curves for a particular FET.

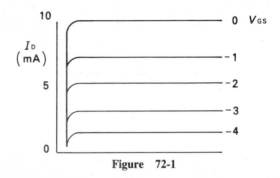

Figure   72-1

Note that with V_{GS} at 0, the FET is already in a state of conduction. If we desired an idle level of 5 mA we would insert an appropriate value of R_S that would set up a V_{GS} of -2 V. The transistor would thus be self-biased. The self-biasing circuit is also self-regulating. If you were to replace the FET in such a configuration with one that had a higher amplifying capability, the increased output current would result in a more negative value of V_{GS}, which in turn would limit the gain.

EXAMPLE 72-1

By referring to Fig. 72-2, determine the required value of R_S for biasing the transistor at the 3-mA level if $V_{RS} = -3$ V. (R_G is usually a 1-MΩ resistance.)

$$R_s = -\frac{V_{RS}}{I_D} = -\frac{(-3\text{ V})}{3\text{ mA}} = 1\text{ k}\Omega$$

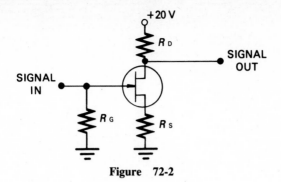

Figure 72-2

EXAMPLE 72-2

If we use the curves in Fig. 72-1, we find that a -2-V source bias will turn the transistor of Fig. 72-2 half on, resulting in 10 V across the transistor. Since we have set up 2 V across R_S, then the remaining 8 V must be across R_D. Calculate R_D.

$$R_D = \frac{V_{RD}}{I_D} = \frac{8\,V}{5\,mA} = 1600\ \Omega$$

EQUATION 73		
BJT SATURATION		
$$I_{SAT} = \dfrac{V_{CC}}{R_C}$$	I_{SAT} = saturation current	A
	V_{CC} = supply voltage	V
	R_C = collector resistor	Ω

The circuit of Fig. 73-1 is a basic one-transistor amplifier. Voltage V_{CC} sets up a current through the base-emitter junction and turns the transistor on to some degree. AS R_B is decreased, this turning on process continues until complete saturation occurs, at which time the collector-emitter portion appears as a closed switch. At this point, the only factor limiting the main flow of current is the collector resistor itself. Applying Equation 73 to Fig. 73-1, you will get a saturation current of 20 mA (20 V/1 kΩ). If you continue to force more current into the base by lowering R_B, the collector current will never exceed the saturation level of 20 mA. Once the unit is on, it is on.

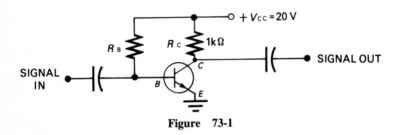

Figure   73-1

One of the more useful applications of Equation 73 is setting the transistor for class A amplification. This will result in nondistorted reproduction. Once you have derived I_{SAT}, simply take half that value and select the proper base resistor to produce it. Use Equation 62 to make this calculation.

EXAMPLE 73-1

Regardless of how much current we force through the base junction, what is the maximum current that can flow through the load in Fig. 73-2?

$$I_{SAT} = \frac{V_{cc}}{R_c} = \frac{30}{100} = 300 \text{ mA}$$

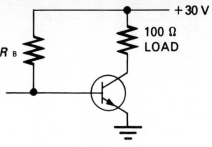

Figure   73-2

EXAMPLE 73-2

If the transistor in Fig. 73-3 has a beta of 100, what should the value of R_B be for class A amplifier operation?

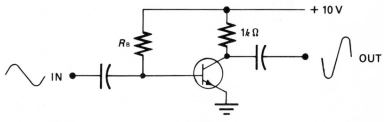

Figure   73-3

First find the saturation current and divide it by 2:

$$I_{SAT} = \frac{V_{CC}}{R_C} = \frac{10}{1000} = 10 \text{ mA} \qquad \tfrac{1}{2}I_{SAT} = 5 \text{ mA}$$

Now find the base current to produce this value by using Equation 62:

$$I_B = \frac{I_C}{\beta} = \frac{5 \times 10^{-3}}{100} = 50 \text{ }\mu\text{A}$$

Finally, find the value of the base resistor:

$$R_B = \frac{V_{CC} - V_{BE}}{I_B} \qquad (V_{BE} = 0.7 \text{ V for a silicon transistor})$$

$$= \frac{10 - 0.7}{50 \times 10^{-6}}$$

$$= 186 \text{ k}\Omega$$

<table>
<tr><td colspan="2" align="center"># EQUATION 74
RIPPLE FACTOR</td></tr>
</table>

$$RF = \frac{\textbf{rms ripple}}{V_{DC}}$$

RF = ripple factor
rms ripple = rms ripple voltage   V
V_{DC} = average DC supply voltage   V

The primary purpose of any given DC power supply is to provide a DC voltage to operate another circuit. An absolutely pure DC output would have a ripple factor of 0. This factor may be converted to percent ripple by multiplying by 100—in this case, 0 percent. The output from a more typical supply with a ripple factor of 0.044, or 4.4 percent, is shown in Fig. 74-1. Note that to obtain the RF we must first convert the 1 V p-p ripple into volts rms before dividing by the average 8-V value. The amount of ripple shown above may be tolerated by some circuits but not others. A high-gain audio amplifier would, for instance, convert this amount of ripple into audible hum. A digital computer might interpret this much ripple as data and produce errors in calculations. To remedy these problems, a larger filter capacitor should be added to the existing one to reduce the ripple factor.

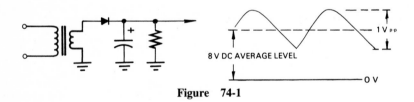

Figure   74-1

EXAMPLE 74-1

Calculate the ripple factor of the power supply shown in Fig. 74-2. First determine the rms ripple:

$$\textbf{RMS ripple} = 0.707 \times \frac{\textbf{p-p ripple}}{2} = 1.77V$$

Now find the ripple factor:

$$RF = \frac{\textbf{rms ripple}}{V_{DC}} = \frac{1.77}{9} = 0.197$$

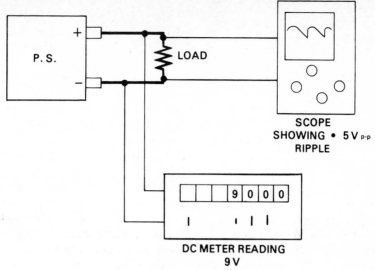

Figure   74-2

EXAMPLE 74-2

Convert the answer in Example 74-1 to percent ripple:

Percent ripple = RF × 100 = 0.197 × 100 = 19.7 percent

EQUATION 75
FILTER CAPACITANCE

$$C = \frac{I \times T}{V}$$

C = filter capacitance	F
I = load current	A
T = ripple time	s
V = p-p ripple	V

Almost every piece of electronic equipment requires some form of power supply to drive it. Equation 75 allows the determination of the exact size of filter capacitor needed in the supply. The prime entry is the amount of current that will be drawn by the device to be driven. This current, in turn, is what determines the value of capacitance required. The T factor is also important and relates to the type of supply being employed. Enter a value of 16.6 ms if you are using a half-wave supply and 8.3 ms if you are using a full-wave unit. These times are derived from the 60-Hz line frequency and the type of rectification. The last entry in the equation is V. It is the actual peak-to-peak ripple that can be tolerated from the supply. Equation 74 may be employed to determine this value. A typical level for a 10-V supply is approximately 0.25 V.

Equation 75 brings out the very important fact that full-wave rectification requires only one-half the filter capacity required by half-wave rectification.

EXAMPLE 75-1

We require a full-wave power supply to drive a 200 mA load with 10 V DC. The rms ripple is to be no greater than 0.25 V. What size should the filter capacitor be? See Fig. 75-1.

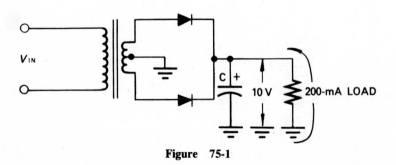

V_{IN} C +   10 V   200-mA LOAD

Figure   75-1

First convert the rms ripple to peak-to-peak.

P-p ripple $= 2 \times 1.414 \times$ **rms ripple** $= 2.828 \times 0.25 = 0.707$ **V**

Then solve for C:

$$C = \frac{I \times T}{V}$$

$$= \frac{(200 \times 10^{-3}) \times (8.3 \times 10^{-3})}{0.707}$$

$$= \frac{1.66 \times 10^{-3}}{0.707}$$

$$= 2348 \; \mu\text{F}$$

The next higher commercial value is 3000 μF.

EXAMPLE 75-2

In Fig. 75-1, if the load is increased to 300 mA, what value should C be changed to?

$$C = \frac{I \times T}{V}$$

$$= \frac{(300 \times 10^{-3}) \times (8.3 \times 10^{-3})}{0.707}$$

$$= \frac{2.49 \times 10^{-3}}{0.707}$$

$$= 3522 \; \mu\text{F}$$

The higher commercial unit is 4000 μF.

EQUATION 76
TRANSISTOR DRIVER

$$R_B = \frac{V_{\text{IN}} - V_{BE}}{2 \times I_B}$$

R_B = base resistance	Ω
V_{IN} = base drive voltage	V
V_{BE} = base-to-emitter voltage	V
I_B = base current	A

Quite often it is required that a very low-level current drives a heavier-current device. A BJT as a current-amplifying device will fulfill this function. The load to be driven can be any device from a lamp to a relay. The transistor, being a current amplifier, will utilize a small base current as a trigger and switch the load, which will be in the collector portion of the circuit. Higher-beta transistors require less base current to turn on than do lower-beta ones. Equation 76 determines the correct value of series base resistance to be used. A high enough resistance must be employed so as not to overdrive the base, thus damaging it. At the same time, the resistance must be low enough to *ensure* turnon. The multiplier of 2 in the denominator of the equation will ensure this. The value of V_{BE} will be 0.7 V for silicon transistors and 0.3 V for germanium ones. You can calculate I_B itself by using Equation 62.

EXAMPLE 76-1

Design a transistor circuit that will activate a 12-V, 250-mA lamp when a 2-V signal is present, as in Fig. 76-1. The transistor's beta is 100.

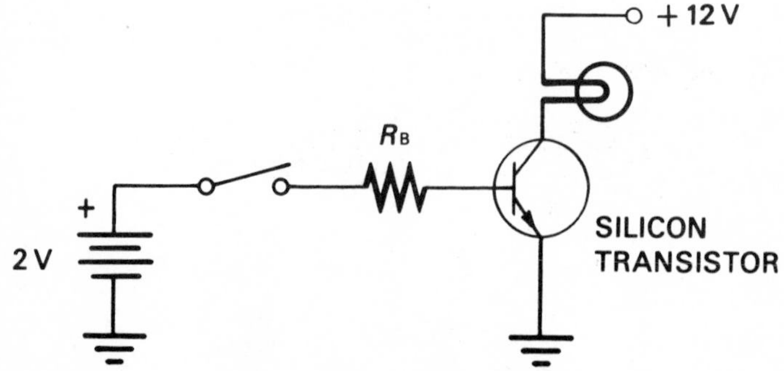

Figure 76-1

Using Equation 62, find the necessary base current:

$$I_B = \frac{I_C}{\beta} = \frac{250 \times 10^{-3}}{100} = 2.5 \text{ mA}$$

Now calculate the base resistance, using Equation 76:

$$R_B = \frac{V_{IN} - V_{BE}}{2 \times I_B} = \frac{2 - 0.7}{2 \times (2.5 \times 10^{-3})} = 260 \ \Omega$$

EXAMPLE 76-2

Design a 6-V relay driver with a protection diode. The relay draws 50 mA, the transistor's beta is 150, and the drive signal is 3 V. (See Fig. 76-2.)

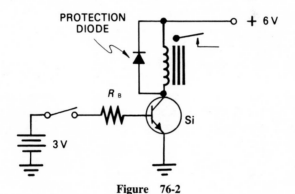

Figure 76-2

First find the necessary base current:

$$I_B = \frac{I_C}{\beta} = 333 \ \mu A$$

Now find the value of R_B:

$$R_B = \frac{V_{IN} - V_{BE}}{2 \times I_B} = \frac{3 - 0.7}{2 \times (333 \times 10^{-6})} = 3448 \ \Omega$$

EQUATION 77
OUTPUT IMPEDANCE

$$\boldsymbol{Z_o} = \boldsymbol{R_C} \,\|\, \boldsymbol{R_L}$$

Z_o = output impedance	Ω
R_C = collector resistor	Ω
R_L = load resistance	Ω

The total load that any transistor "works into" is one of the prime factors in determining stage gain. This total output load makes up the total output impedance.

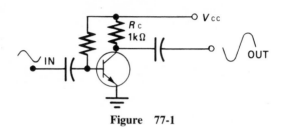

Figure 77-1

In Fig. 77-1 the output impedance is simply the collector resistor (1 kΩ), if the output terminal is left "floating." In Fig. 77-2, where an actual load exists, if we trace the current paths from the collector to ground, two exist, one through R_C and the other through R_L. Since a typical power supply will have very high amounts of capacitance, all signals will pass through R_C and the supply to ground. Therefore, R_C and R_L are in parallel and together make up the total output impedance into which the transistor must work. The output coupling capacitor will have a sufficiently low X_C that it need not be considered in this analysis.

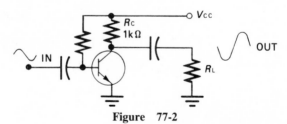

Figure 77-2

EXAMPLE 77-1

Determine the total output impedance of the single-stage amplifier in Fig. 77-3.

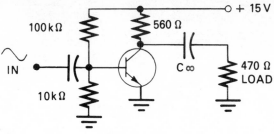

Figure   77-3

$Z_o = R_C \| R_L$ (which reads, "R_C paralleled with R_L")

$= 560\ \Omega \| 470\ \Omega$

$= \dfrac{560 \times 470}{560 + 470} = 255.5\ \Omega$

EXAMPLE 77-2

Determine the total output impedance of the first stage of Fig. 77-4.

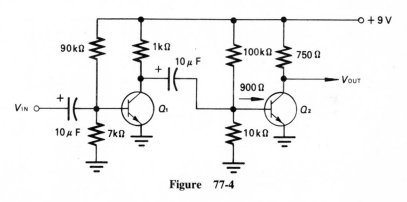

Figure   77-4

Here, R_L comprises 100 kΩ, 10 kΩ, and 900 Ω of input impedance to Q_2.

$$100\ k\Omega \| 10\ k\Omega \| 900\ \Omega = 819\ \Omega$$

$$Z_o = R_C \| R_L = 1\ k\Omega \| 819\ \Omega = 450\ \Omega$$

EQUATION 78
TRANSISTOR STAGE GAIN

$$A_V = \frac{R_C \parallel R_L}{R_E}$$

A_V = voltage gain

R_C = collector resistor Ω

R_L = load resistance Ω

R_E = emitter resistor Ω

The voltage gain of a single-stage amplifier is the degree to which an input signal will be amplified at the output. The primary factor that determines this degree is the ratio of emitter resistance to the collector's load, if the configuration is that shown in Fig. 78-1. Lowering the value of R_E will increase A_V, to the point at which the emitter is directly grounded. In this case the only remaining emitter resistance is that of the internal emitter junction itself. This resistance may be roughly calculated by $r_{ej} \approx 0.03/I_C$ where 0.03 is a constant and I_C is the maximum collector current that can flow. When the emitter resistor is greater than 100 Ω, the internal emitter junction resistance may be ignored.

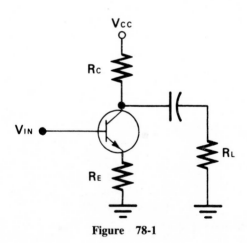

Figure 78-1

You might ask, Why not ground the emitter in all cases to obtain maximum gain? With the emitter directly grounded, the transistor is very susceptible to current shifting owing to thermal drift. Therefore, we sacrifice a small amount of gain for stability by inserting some emitter resistance.

EXAMPLE 78-1

Calculate the voltage gain and determine the amplitude of the output signal at the load in Fig. 78-2 for an input signal of 300 mV.

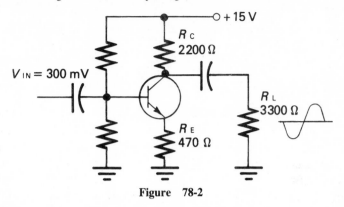

Figure 78-2

First, from Equation 78, find A_V:

$$A_V = \frac{R_C \| R_L}{R_E} = \frac{2200 \| 3300}{470} = 2.8$$

Now determine the output signal voltage:

$$V_{OUT} = 2.8 \times 300 \text{ mV} = 840 \text{ mV}$$

NOTE The minus sign indicates a phase shift of 180°.

EXAMPLE 78-2

Calculate the voltage gain of the common-emitter amplifier of Fig. 78-3.

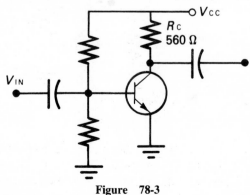

Figure 78-3

First determine the maximum collector current:

$$I_c = \frac{V_{CC}}{R_C} = \frac{10}{560} = 17.9\,\text{mA}$$

Since there is no emitter resistor, r_{ej} must be used:

$$r_{ej} - \frac{0.03}{I_C} - \frac{0.03}{17.9 \times 10^{-3}} - 2\,\Omega$$

Then A_V can be calculated:

$$A_V = \frac{R_C}{r_{ej}} = \frac{560}{2} = 280$$

EQUATION 79
OP AMP GAIN

$$A_V = \frac{R_F}{R_i}$$
A_V = voltage gain
R_i = input resistor $\quad \Omega$
R_F = feedback resistor $\quad \Omega$

Outstanding amplifier performance may be obtained from any of the currently available operational amplifiers (op amps). Using the inverting ($-$) input as the signal input terminal produces an amplified, inverted output, as shown in Fig. 79-1.

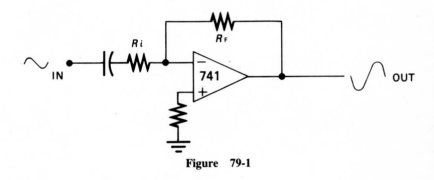

Figure   79-1

The resistance R_i will determine the input impedance to the amplifier. Its value should be in the range of 2.2 to 22 kΩ. The relationship between this resistance and R_F, the feedback resistor, establishes the overall gain of the system. R_F should not exceed 1 MΩ, as at this level instability may result. The resistance from the noninverting ($+$) input to ground is $R_i \parallel R_F$. When you use a single-ending power supply, half of V_{CC} should be applied to the noninverting terminal from a 50/50 voltage divider. R_F may be replaced by a potentiometer in order to vary the gain of the stage.

EXAMPLE 79-1

Calculate the stage gain in Fig. 79-2.

$$A_V = \frac{R_F}{R_i} = \frac{22,000}{2200} = 10$$

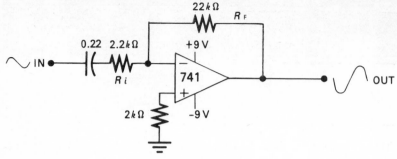

Figure 79-2

EXAMPLE 79-2

If you are using one-quarter of a type 324 op amp and R_i is 4.7 kΩ and R_F is 100 kΩ, determine the stage gain and draw the schematic.

$$A_V = \frac{R_F}{R_i} = \frac{100 \times 10^3}{4700} = 21.3$$

The schematic is shown in Fig. 79-3. The voltage divider on the noninverting input supplies ½ V_{CC} to the input.

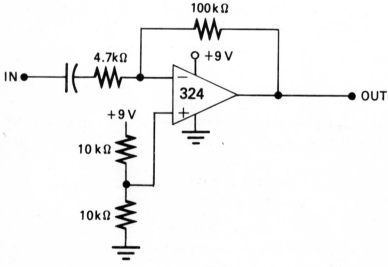

Figure 79-3

EQUATION 80 MONOSTABLE TIME		
$T = 1.1 \times RC$	T = time of pulse	s
	R = resistance	Ω
	C = capacitance	F

The output from a monostable multivibrator is a single-shot, one-time event. When triggered, the monostable produces a low-to-high transition at its output. The time for which this pulse remains high before falling back to 0 may be calculated with Equation 80. It must be stressed that this particular equation is applicable only to the 555 integrated monostable circuit of Fig. 80-1. As shown in the figure, a negative-going pulse at the input (pin 2) is required to trigger the unit. The values of R and C set the length of time that the output will remain high. The general practice is to select the time required and a value for C and then to solve for $R = I/(1.1 \times C)$.

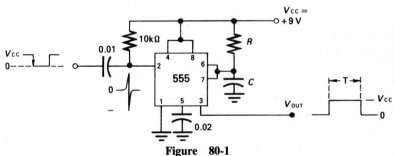

Figure 80-1

EXAMPLE 80-1

A monostable is constructed with R = 10 kΩ and C = 0.22 μF, as in Fig. 80-1. What is the pulse duration at pin 3?

$$T = 1.1 \times RC$$
$$= 1.1 \times (10 \times 10^3) \times (0.22 \times 10^{-6})$$
$$= 2.42 \text{ ms}$$

EXAMPLE 80-2

If a pulse duration of 650 μs is required and a 0.001-μF capacitor has been selected, what value must R be?

$$R = \frac{I}{1.1 \times C} = \frac{650 \times 10^{-6}}{1.1 \times (0.001 \times 10^{-6})} = \frac{650 \times 10^{-6}}{1.1 \times 10^{-9}}$$
$$= 591 \text{ k}\Omega$$

EQUATION 81 ASTABLE FREQUENCY		
$f = \dfrac{1.44}{(R_A + 2R_B)C}$	f = oscillator frequency	Hz
	R_A = frequency-timing resistor A	Ω
	R_B = frequency-timing resistor B	Ω
	C = timing capacitor	F

Equation 81 applies only to the 555 integrated circuit when used as an astable multivibrator. The typical astable circuit is shown in Fig. 81-1, where the output is a square wave. The upper frequency limit of the 555 IC is about 500 kHz. Resistors R_A and R_B, along with capacitor C, determine the frequency of the output at pin 3. Capacitor C charges through R_A and R_B and discharges through R_B only. The duty cycle of the output may be varied by adjusting the ratio of these two resistances.

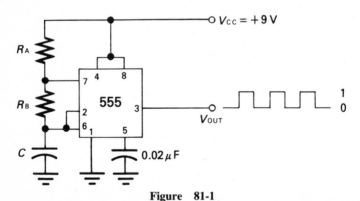

Figure 81-1

In summary: Equation 81 is utilized in determining the free-running frequency and the formula $D = R_B/(R_A + 2R_B)$ is employed in determining the duty cycle. For thermal stability, resistor R_A should be no lower than 2 kΩ. The output load from pin 3 to ground should be no less than 50 Ω.

EXAMPLE 81-1

When $R_A = 3.3$ kΩ, $R_B = 10$ kΩ, and $C = 0.05$ μF, what is the output frequency?

$$f = \frac{1.44}{(R_A + 2R_B)C} = \frac{1.44}{(3.3 \times 10^3 + 20 \times 10^3)(0.05 \times 10^{-6})} = 1236 \text{ Hz}$$

EXAMPLE 81-2

What is the range of the variable-frequency circuit of Fig. 81-2?

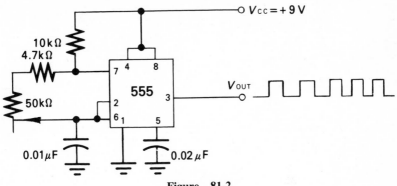

Figure 81-2

Upper frequency
(Potentiometer closed)

$$f = \frac{1.44}{(10 \times 10^3 + 9.4 \times 10^3)(0.01 \times 10^{-6})}$$

$$= 7423 \text{ Hz}$$

Lower frequency
(Potentiometer open)

$$f = \frac{1.44}{(10 \times 10^3 + 10.94 \times 10^4)(0.01 \times 10^{-6})}$$

$$= 1206 \text{ Hz}$$

EQUATION 82			
VCO FREQUENCY			

$$f = \frac{2(V_{CC} - V_c)}{R \times C \times V_{CC}}$$

f = oscillator frequency	Hz
V_{CC} = supply voltage	V
V_c = control voltage	V
R = timing resistor	Ω
C = timing capacitor	F

Equation 82 is applicable only to the 566 voltage-controlled oscillator (VCO) integrated circuit (Fig. 82-1). The oscillator's frequency is set primarily by R and C; R should be in the range of 2 to 20 kΩ. The frequency is *also* controlled by the voltage at pin 5, which in the equation is designated V_c. Note that both triangular and square waveforms are available from pins 4 and 3, respectively.

In the schematic, voltage V_c is obtained from a voltage divider (R_x and R_y), which places 10.4 V on pin 5. This control voltage may come from an external source, such as a modulator, and may be variable. Voltage V_c should be in the range of 3 to 11 V when V_{CC} = 12 V. This range will yield a 10:1 frequency sweep range.

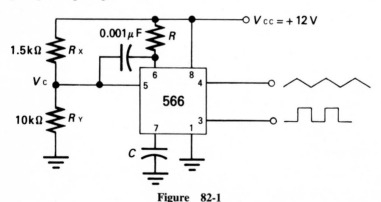

Figure 82-1

EXAMPLE 82-1

Referring to Fig. 82-1, assume V_c = 10.4 V, V_{CC} = 12 V, R = 5.6 kΩ, and C = 0.22 μF. What is the VCO frequency?

$$f \cong \frac{2(V_{CC} - V_c)}{R \times C \times V_{CC}} \cong \frac{2(12 - 10.4)}{(5.6 \times 10^3) \times (0.22 \times 10^{-6}) \times 12} \cong \frac{3.2}{14.8 \times 10^{-3}} \cong 217 \text{ Hz}$$

EXAMPLE 82-2

Assume in Example 82-1 that R_X and R_Y are changed to 4.7 and 10 kΩ, respectively, thus changing V_C. What is the new VCO output frequency?

First determine the new V_C:

$$V_c = \frac{R_Y}{R_X + R_Y} \times V_{CC}$$

$$= \frac{10 \times 10^3}{4.7 \times 10^3 + 10 \times 10^3} \times 12$$

$$= 8.16 \text{ V}$$

Now determine the frequency:

$$f \cong \frac{2(V_{CC} - V_c)}{R \times C \times V_{CC}} \cong \frac{2(12 - 8.16)}{(5.6 \times 10^3) \times (0.22 \times 10^{-6}) \times 12} \cong \frac{7.68}{14.8 \times 10^{-3}} \cong 519 \text{ Hz}$$

EQUATION 83
CAPACITOR CHARGE SHIFT

$$v = V(1 - \epsilon^{-t/RC})$$

v = voltage change   V
V = driving voltage   V
t = time   s
RC = time constant   s

Equation 28 pointed out that a capacitor will reach full charge after five time constants have elapsed. Quite often, however, we need to know the charge across the capacitor at a time less than $5RC$. Equation 83 solves for v, which is the change that takes place in a time interval t. The value of R is the Thévenin resistance as seen by the capacitor (see Equation 13).

In Fig. 83-1a, R equals 10 kΩ. In Fig. 83-1b, R would be kΩ. Thévenize by imagining a short circuit in place of the source V. In the equation, the factor e is 2.72, the base for the natural logarithm. When working these problems, first divide $-t$ by RC; this division will yield a negative value. At this point, employ the e^x function on your calculator.

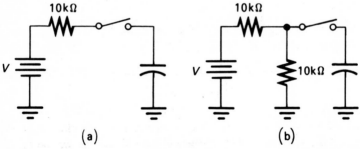

(a)   (b)

Figure 83-1

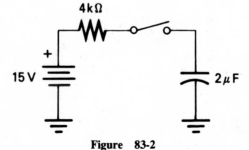

Figure 83-2

EXAMPLE 83-1

In Fig. 83-2, find v across the capacitor 12 ms after the switch is closed. First determine the RC time constant:

$$RC = (4 \times 10^3) \times (2 \times 10^{-6}) = 8 \times 10^3 s$$

Now determine v at 12 ms:

$$v = V(1 - e^{-t/RC})$$
$$= 15(1 - e^{(-12 \times 10^{-3})/(8 \times 10^{-3})})$$
$$= 15(1 - e^{-1.5}) = 15(1 - 0.223) = 15(0.777)$$
$$= 11.66 \text{ V}$$

EXAMPLE 83-2

In Fig. 83-3, find v across the capacitor 150 μs after the switch is closed.

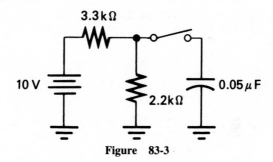

Figure 83-3

First, when we Thévenize the circuit according to Equation 13, we find that $V = 4$ V and $R = 1320$ Ω. Therefore, RC is 66 μs and we can calculate v as follows:

$$v = V(1 - e^{-t/RC})$$
$$= 4(1 - e^{(-150 \times 10^{-6})/(66 \times 10^{-6})}) = 4(1 - e^{-2.27})$$
$$= 4(1 - 0.103) = 4(0.897)$$
$$= 3.59 \text{ V}$$

EQUATION 84
CAPACITOR CHARGE TRANSITION TIME

$$t = -RC \ln \left(1 - \frac{v}{V} \right)$$

t = transition time   s
RC = time constant   s
v = voltage shift   V
V = driving voltage   V

It is physically impossible for a capacitor to shift from one voltage level to another instantly. Some transition time is required. Equation 84 relates directly to Equation 83; however, instead of solving for voltage shift, it determines this time period. As you can see by examining the formula, this length of time is related to both the RC time constant and the voltage shift. Also incorporated into the formula is the natural logarithm function, upon which all capacitor charge and discharge rates are based. The value of v simply indicates the amount of voltage shift and does not necessarily have to begin at 0. An example is a capacitor having a 10-V charge which is charged further to a 16-V level when a charging switch is closed. In this case v equals 6 V—the amount of change. Note that Equations 83 and 84 apply to discharge voltages and times as well.

EXAMPLE 84-1

In Fig. 84-1, how long after the charging switch is closed will the potential across the capacitor be 11.66 V if the capacitor is being charged from 0?

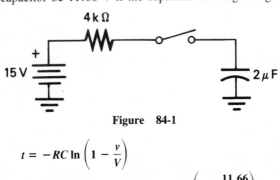

Figure 84-1

$$t = -RC \ln \left(1 - \frac{v}{V} \right)$$

$$= -(4 \times 10^3) \times (2 \times 10^{-6}) \times \ln \left(1 - \frac{11.66}{15} \right)$$

$$= -8 \times 10^{-3} \times \ln 0.22267$$

$$= -8 \times 10^{-3} \times (-1.5)$$

$$= 12 \text{ ms}$$

EXAMPLE 84-2

If $V = 10$ V, $R = 1320$ Ω, $C = 0.05$ μF, and $v = 8.97$ V, find t.

$$t = -RC \ln \left(1 - \frac{v}{V} \right)$$

$$= -1320 \times (0.05 \times 10^{-6}) \ln \left(1 - \frac{8.97}{10} \right)$$

$$= -1320 \times (0.05 \times 10^{-6}) \ln 0.103$$

$$= -1320 \times (0.05 \times 10^{-6})(-2.27)$$

$$= 150 \text{ μs}$$

EQUATION 85
DECIBELS

$$N_{dB} = 10 \times \log \frac{P_{OUT}}{P_{IN}}$$

N_{dB} = number of decibels   dB
P_{OUT} = power out   W
P_{IN} = power in   W

The human ear responds to the differences between sound levels on a logarithmic (base 10) scale. That is, for a 1-kHz tone to be made to seem twice as loud, its power has to be increased 10 times. It is imperative, therefore, that you understand the decibel (dB).

As far as change in sound intensity (loudness) is concerned, the ear can perceive a change of 2.5 dB. For example, if two oscillators produce a 1-kHz tone at 70-dB and 71-dB sound levels, they will seem equal in intensity. Overall sound levels over 80 dB for prolonged periods may be damaging to the ear.

Application of Equation 85 will show that a 100-W amplifier will not be able to produce a sound twice as loud as a 50-W unit for the same 1-W input:

$$N_{dB} = 10 \log \frac{100 \text{ W}}{1 \text{ W}} \qquad N_{dB} = 10 \log \frac{50 \text{ W}}{1 \text{ W}}$$

$$= 20 \text{ dB} \qquad\qquad\quad = 17 \text{ dB}$$

In fact, it takes twice the power to produce a sound just perceptibly louder. In other words, with a 3-dB difference between the two, one would be louder but not by much. Equation 85, then, yields the change in dB between input and output power levels.

We may also employ the standard reference level of 1 mW into 600 Ω for P_{IN} as the 0-dB level. The answer in this case will be written as so many dB above 1 mW and represents the actual power level in a circuit. Levels referred to this way are often written in dBm, or in volume units (VU).

EXAMPLE 85-1

What is the gain in dB of an amplifier if an input of 500 mW produces a 20-W output?

$$N_{dB} = 10 \times \log \frac{P_{OUT}}{P_{IN}}$$

$$= 10 \times \log \frac{20}{0.5} = 10 \times \log 40$$

$$= 10 \times 1.6 = 16 \text{ dB}$$

EXAMPLE 85-2

Find the dBm gain of a microphone whose output is 0.000008W.

$$N_{dBm} = 10 \times \log \frac{P_{OUT}}{0.001}$$

$$= 10 \times \log \frac{8 \times 10^{-6}}{0.001} = 10 \times \log 0.008$$

$$= 10 \times (-2) = -20\,dBm$$

Thus, the output of the microphone is attenuated 20 dB below the reference level of 1 mW into 600 Ω.

EQUATION 86
RESONANT COMPONENT VALUES

$$L \cong \frac{1400}{2\pi f}$$

L =	inductance	H
2π =	a constant	
f =	frequency	Hz

Equation 86 and its complement equation for finding capacitance, ($C - 1/(2\pi f \times 1400)$), are used in selecting proper values at resonance. Both are unique to this text in that they are the author's derivation. While it is true that when you construct a resonant tank circuit, any two values of L and C may be matched to obtain the required resonant frequency, often the selection is impractical. On occasion, a very small capacitance is used with a very large inductance and the circuit performance is less than desired, whereas if medium-sized components were employed the performance would improve. These equations give the designer a starting point at which both components have a reactance of 1400 Ω at resonance. The use of Equation 86 will not result in perfection, but it is far better than selecting components at random. Coupling Equations 86 and 87 will allow you to effectively select components and construct efficient tuned circuits.

EXAMPLE 86-1

What value of L should be used in the tank circuit of the RF amplifier in Fig. 86-1?

$$L \cong \frac{1400}{2\pi f} \cong \frac{1400}{6.28 \times (27 \times 10^6)}$$

$$\cong \frac{1400}{1.7 \times 10^8} \cong 8.26\ \mu\text{H}$$

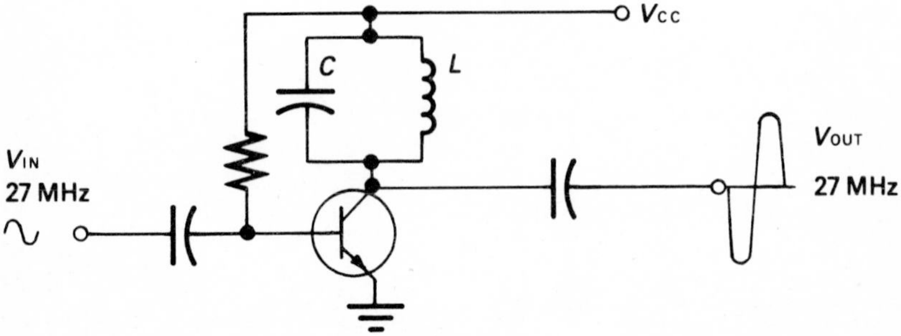

Figure 86-1

EXAMPLE 86-2

In Example 86-1, what value of capacitance should be employed?

$$C \cong \frac{1}{2\pi f \times 1400} \cong \frac{1}{6.28 \times (27 \times 10^6) \times 1400}$$

$$\cong \frac{1}{2.37 \times 10^{11}} \cong 4.2 \ \mu F$$

As a double check:

$$f_0 = \frac{1}{2\pi \sqrt{LC}}$$

$$= \frac{1}{2\pi \sqrt{(8.3 \times 10^{-6}) \times (4.2 \times 10^{-12})}}$$

$$= 27 \ \text{MHz}$$

EQUATION 87
COIL CONSTRUCTION

$$N = \frac{\sqrt{L(9r + 10l)}}{r}$$

N = number of turns
L = inductance   μH
r = radius   in
l = length   in

The required number of turns for an air-core coil may be determined by Equation 87. You may first employ Equation 86 to find the required value of inductance. Once this is known, it is then only necessary to select a coil form of given diameter and to wind the coil itself.

This equation is applicable only to single-layered windings. Fig. 87-1 shows the designations of r and l. The copper wire windings have been cut away to show that the radius r is actually taken to the center of the wire. For frequencies up to 7 MHz use #26 wire; from 7 to 30 MHz use #24 or #22. For higher frequencies, use self-supporting #20 or #18 without a form. Lower frequency forms may be phenolic, paper, or ceramic. Ceramic types provide the highest degree of stability, phenolic the next highest, and paper the lowest.

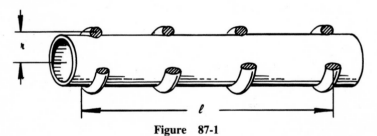

Figure 87-1

EXAMPLE 87-1

If we are using a ½-in diameter plastic form and intend to wind a coil that is ¾ in long, how many turns will be required to produce a 15-μH coil?

If the diameter is 0.5 in, then the radius is 0.25 in. Then,

$$N = \frac{\sqrt{L(9r + 10l)}}{r}$$

$$= \frac{\sqrt{15(9 \times 0.25 + 10 \times 0.75)}}{0.25}$$

$$= \textbf{48 turns}$$

EXAMPLE 87-2

At what resonant frequency should the coil in Example 87-1 be employed and what gage wire should be used?

Since $X_L = 1400 \ \Omega$ at f_0 and $X_L = 2\pi fL$:

$$f = \frac{X_L}{2\pi L} = \frac{1400}{6.28 \times (15 \times 10^{-6})}$$

$$= 14.9 \ \text{MHz}$$

The correct gage of wire is either #24 or #22.

EQUATION 88
CAPACITIVE CONNECTIONS

$C = \dfrac{1 \times 10^6}{2\pi f(R \times 0.1)}$	C = capacitance	µF
	R = circuit resistance	Ω
	f = lowest frequency	Hz

Capacitors are used in electronic circuits for one of two main purposes: coupling or bypass applications. We shall discuss each separately, although both are directly related.

In Fig. 88-1a, the coupling capacitor feeds the signal from one stage to another. For maximum energy transfer its reactance should be one-tenth that of the input impedance of the following stage. This must be true for the lowest frequency to be passed.

Bypass capacitors pass signals around resistances to ground, as in Fig. 88-1b. They should have a reactance at the lowest frequency to be bypassed that is one-tenth the circuit resistance to be bypassed. Bypass capacitors are often employed to cut off higher frequencies while not affecting lower ones. A good example of this is in the rejection of (radio-frequency) RF signals (by bypassing them to ground) in audio amplifiers.

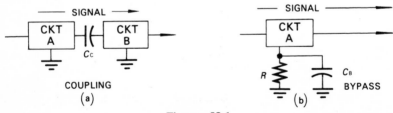

Figure 88-1

EXAMPLE 88-1

Signals from 100 Hz to 15 kHz are to be amplified by the operational amplifier of Fig. 88-2. Its input impedance is determined by R_i. Find the value of C_C.

$$C = \frac{1 \times 10^6}{2\pi f(R \times 0.1)} = \frac{1 \times 10^6}{6.28 \times 100 \times (10 \times 10^3 \times 0.1)}$$

$$= 1.6 \ \mu F$$

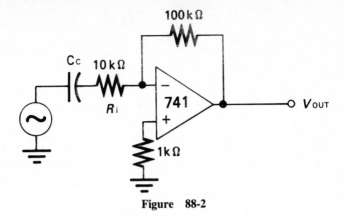

Figure 88-2

EXAMPLE 88-2

Calculate the value of the bypass capacitor for the single-stage amplifier of Fig. 88-3 if $f_{LOW} = 200$ Hz.

$$C = \frac{1 \times 10^6}{2\pi f(R \times 0.1)}$$
$$= \frac{1 \times 10^6}{6.28 \times 200 (470 \times 0.1)} = 17\ \mu F$$

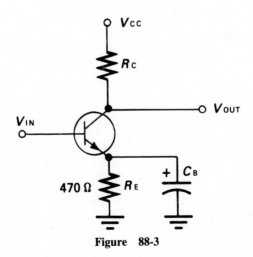

Figure 88-3

EQUATION 89
BAND-PASS AMPLIFIER

$$R_1 = \frac{2}{2\pi \times \Delta f \times C}$$

R_1 = feedback resistance Ω
2π = a constant
Δf = bandwidth   Hz
C = capacitance   F

Outstanding bandpass amplifiers that are either narrowband or broadband may be designed by using Equation 89 along with the two equations listed below. The rolloff is sharp, precise, and within the complete control of the designer. (See Fig. 89-1.)

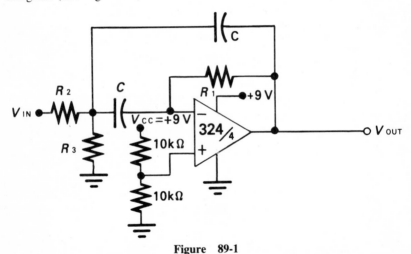

Figure 89-1

Start by selecting a value for C from Table 89-1 (both capacitors will be the same). Use Equation 89 for determining the value of R_1. Once it has been found, use the equation $R_2 = R_1/(2 \times A_v)$, where A_v is the desired gain (usually from 2 to 10). Then decide upon the Q factor required and solve for $R_3 = R_1/(4 Q^2 - 2A_v)$. It should be remembered from Equation 56 that Q is the center frequency divided by the bandwidth required at the -3-dB points.

EXAMPLE 89-1

Find R_1, R_2, and R_3 for a band-pass amplifier where A_V is 2, Q is 1.5, and the center frequency is to be 800 Hz.

If Q = 1.5, then

$$BW = \Delta f = \frac{f_0}{Q} = \frac{800}{1.5} = 533 \, Hz$$

Select $C = 0.022 \, \mu F$ (see Table 89-1). Then,

$$R_1 = \frac{2}{2\pi \times (533 \times .02 \times 10^{-6})} = 30 \, k\Omega$$

$$R_2 = \frac{R_1}{2 \times A_V} = \frac{30 \times 10^3}{2 \times 2} = 7500 \, \Omega$$

$$R_3 = \frac{R_1}{4Q^2 - 2A_V} = \frac{30 \times 10^3}{4(1.5)^2 - 4} = 6 \, k\Omega$$

Table 89-1 Capacitor Selection

f_0	C
100 Hz	0.22 μF
1 kHz	0.022 μF
10 kHz	0.0022 μF

COMMUNICATIONS

EQUATION 90
AM PERCENT MODULATION

$$\%M = \frac{V_M}{V_C} \times 100$$

$\%M$ = percent of amplitude modulation
V_M = peak modulating voltage   V
V_C = peak carrier voltage   V

If not modulated with an audio signal, the output of an RF amplifier is at a constant amplitude. Reception of such a signal on an AM receiver will be indicated by the presence of a "dead air space." V_C in Equation 90 is the peak voltage of this unmodulated carrier. If an audio signal is now superimposed on the main supply line feeding the RF amplifier, the amplifier's output will vary in accordance with this audio signal. When the amplitude of the modulating signal is equal in peak voltage to that of the carrier, 100 percent modulation will have been reached. This is illustrated in Fig. 90-1.

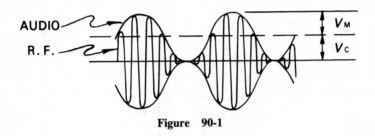

Figure 90-1

In this case, if the carrier voltage is 500 V and the modulating voltage is 500 V:

$$\%M = \frac{V_M}{V_C} \times 100 = \frac{500}{500} \times 100 = 100\%$$

Modulation beyond this point results in distortion from sideband splatter.

EXAMPLE 90-1

A 300-V_{PEAK} carrier signal is modulated with a 175-V_{PEAK} audio signal. What is the percent of modulation?

$$\%M = \frac{V_M}{V_C} \times 100 = \frac{175}{300} \times 100 = 58.3\%$$

EXAMPLE 90-2

In the waveform envelope in Fig. 90-2, what percent modulation is evidently being produced (assume $V_C = 100$ V)?

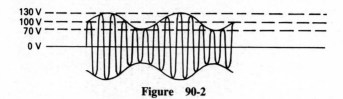

Figure   90-2

The modulating voltage appears to vary between 70 V and 130 V. Thus $V_{\text{p-p}} = 60$ and $V_M = 30$ V. Therefore,

$$\%M = \frac{V_M}{V_C} \times 100 = \frac{30}{100} \times 100 = 30\%$$

EQUATION 91
CRYSTAL SHIFT

$$f_s = -k \times \Delta t \times f$$

f_s = frequency shift	Hz
k = temperature coefficient	ppm/°C
Δt = temperature change	°C
f = crystal frequency	MHz

X-cut transmitting crystals possess a negative temperature coefficient. This means that as temperature goes up, their frequency goes down; as temperature goes down, their frequency goes up. Y-cut crystals may have either a positive or a negative temperature coefficient. A rising temperature on a positive-temperature-coefficient crystal will cause its frequency to increase. When we obtain a crystal from a supplier, we may be given two pieces of information: the frequency at some specified temperature and the temperature coefficient. With this information, and also knowing to what temperature extremes the circuit will be exposed, we can calculate the expected frequency deviation. The temperature change Δt is obtained by subtracting the new operating temperature from the originally specified temperature. Thus, an increasing temperature results in a negative value for Δt. If the results indicate a shift that cannot be tolerated, then a small crystal-heating oven must be employed. The oven will raise the crystal to an exact temperature and hold it there precisely.

EXAMPLE 91-1

An X-cut crystal has a temperature coefficient of -12 ppm/°C and is listed as being 1.5 MHz at 50°C. What will its frequency be if the temperature falls to 30°C?

$$\Delta t = 50 - 30$$
$$= 20°C$$

Therefore:

$$f_s = -k \times \Delta t \times f = -(-12) \times 20 \times 1.5 = 360 \text{ Hz}$$

Since the temperature coefficient is negative and the temperature is decreasing, we must then add the 360 Hz to the original frequency of 1.5 MHz:

$$f + f_s = 1,500,000 + 360 = 1,500,360 \text{ Hz}$$

EXAMPLE 91-2

In Example 91-1, what will the new frequency be if the temperature shifts upward to 60°C?

$$\Delta t = 50 - 60 = -10°C$$

Therefore:

$$f_s = -k \times \Delta t \times f = -(-12) \times (-10) \times 1.5 = -180\,\text{Hz}$$

This shift is negative (to a lower frequency). Thus,

$$f + f_s = 1,500,000 - 180 = 1,499,820\,\text{Hz}$$

EQUATION 92
SIDEBAND POWER

$$P_{SB} = \frac{M^2}{2} \times P_C$$

P_{SB} = sideband power   W
M = degree of modulation
P_C = carrier power   W

Modulating a 1-MHz carrier with an 800-Hz tone will produce sideband signals at 999,200 Hz and 1,000,800 Hz. The actual intelligence of the transmitted signal is in these two sidebands. It is imperative that as much power be pumped into the sidebands as possible to ensure intelligent reception at great distances. M in Equation 92 is the degree of modulation and relates directly to the percent of modulation; e.g., 50 percent modulation equals an M of 0.5 and 30 percent modulation an M of 0.3. The highest this factor can be is 1 at 100 percent modulation. From Equation 92 it is apparent that by increasing the percent of modulation, we are placing more power into the sidebands where it needs to be. Even at 100 percent, however, there will be only *50 percent* of the total carrier power divided between *both sidebands*.

EXAMPLE 92-1

A 50,000-W transmitter is modulated to the 75 percent level by a sine wave. How much power is there in *both* sidebands combined?
NOTE If the percent of modulation is 75 percent, then $M = 0.75$.)

$$P_{SB} = \frac{M^2}{2} \times P_C = \frac{0.75^2}{2} \times 50,000$$
$$= \frac{0.5625}{2} \times 50,000 = 14,063 \text{ W}$$

EXAMPLE 92-2

A 100-W transmitter is modulated to the 100 percent level by a sine wave. How much power is in *each* sideband?
First determine the sideband power:

$$P_{SB} = \frac{M^2}{2} \times P_C = \frac{1^2}{2} \times 100$$
$$= 0.5 \times 100 = 50 \text{ W}$$

Since half the power is in each sideband,

$$P\text{(in one sideband)} = 25 \text{ W}$$

EQUATION 93
FM PERCENT MODULATION

$$\%M = \frac{\Delta f}{75 \times 10^3 \text{ Hz}} \times 100$$

$\%M$ = percent modulation
Δf = carrier frequency swing   Hz

In an AM system the percent of modulation is determined by the amplitude of the modulating signal. The situation in an FM (frequency modulation) system, although similar, is different. Here amplitude changes from the microphone (or other signal source) cause the carrier to shift in frequency. Stronger signals from the microphone cause greater shifts from the center (resting) frequency. The maximum shift allowed by the FCC is 75 kHz. By examining the formula you can see that a 75-kHz swing (Δf) will represent 100 percent modulation. The frequency of the modulating sound will have no effect on percent modulation but will vary the rate at which the carrier frequency changes. Thus, two modulating signals of equal amplitude will both cause a swing of, say, 50 kHz, but if one is of a lower frequency, it will do so at a slower rate. The frequency spectrum for a single FM commercial station is shown in Fig. 93-1.

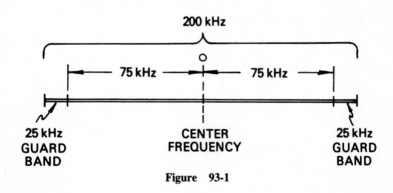

Figure   93-1

EXAMPLE 93-1

An FM transmitter is modulated with a 5000-Hz tone that produces a 50-kHz carrier deviation. What is the percent of modulation?

$$\% M = \frac{\Delta f}{75 \times 10^3} \times 100 = \frac{50 \times 10^3}{75 \times 10^3} \times 100 = 66.7\%$$

EXAMPLE 93-2

If, in Example 93-1, the modulating frequency is increased to 10 kHz, what is the new percent of modulation?

The frequency of the modulating signal has no effect on the percent of modulation, which will remain at 66.7%.

EQUATION 94
PARALLEL TRANSMISSION LINE IMPEDANCE

$Z_0 = 276 \log\left(\dfrac{D}{R}\right)$	Z_0 = characteristic (surge) impedance	Ω
	D = spacing	cm
	R = conductor radius	cm

The primary function of an RF transmission line is to efficiently transfer power from transmitter to antenna or from antenna to receiver. Our main concern here is that the characteristic (surge) impedance of the line match that of the antenna. Even a single conductor possesses inductance, and parallel lines have capacitance as well. This combination will produce an AC impedance $Z_0 = \sqrt{L/C}$. This impedance is a function of both wire spacing and wire diameter, as shown in Fig. 94-1. In actual construction, the twin-lead type of transmission line will have the two conductors separated with some form of plastic spacers or a solid dielectric. Should the antenna Z not match the line Z_0, standing waves will be produced on the line. This will result in a power loss at the antenna. Such a mismatch is usually corrected by changing the impedance of the antenna by pruning it to match the impedance of the fixed line.

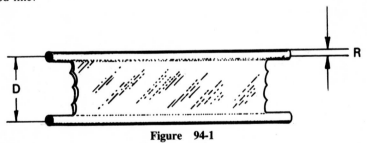

Figure 94-1

EXAMPLE 94-1

Determine the surge impedance of the twin-lead line if each of the wires has a radius of 0.12 cm and the two are spaced 1 cm apart.

$$Z_0 = 276 \log\left(\frac{D}{R}\right) = 276 \log\left(\frac{1}{0.12}\right)$$

$$= 276 \log(8.33) = 276 \times 0.921$$

$$= 254\ \Omega$$

EXAMPLE 94-2

If, in Example 94-1, we increase the spacing to 2 cm, what is Z_0?

$$Z_0 = 276 \log \left(\frac{D}{R} \right) = 276 \log \left(\frac{2}{0.12} \right)$$

$$= 276 \log (16.67) = 276 \times 1.22$$

$$= 337 \ \Omega$$

EQUATION 95
COAXIAL LINE IMPEDANCE

$Z_0 = 138 \log \left(\dfrac{D_S}{D_C} \right)$	Z_0 = characteristic (surge) impedance	Ω
	D_S = sheath diameter	cm
	D_C = conductor diameter	cm

Whereas parallel lines are balanced, coaxial conductors are unbalanced. The outer braided metallic sheath is held at ground potential, while the inner conductor is "hot" (Fig. 95-1). The sheath effectively shields spurious radiation from the line, thus lessening radio-frequency interference (RFI). The characteristic impedance of a coaxial cable is established by the ratio of the diameters of the inner and outer conductors. Commercially available types include RG-8/U at 50 Ω, RG-58A/U at 50 Ω, and RG-59/U at 75 Ω, to mention only a few. Although RG-8/U and RG-58A/U are both rated at 50 Ω, the first is a physically larger type, exhibiting a lower signal loss per foot.

Log assumes base 10. Ln is natural log.

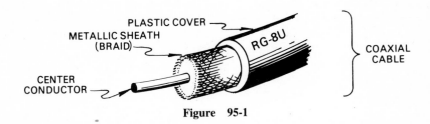

PLASTIC COVER
METALLIC SHEATH (BRAID)
CENTER CONDUCTOR
RG-8U
COAXIAL CABLE

Figure 95-1

EXAMPLE 95-1

Upon measuring a portion of coaxial line we find that the outer metallic sheath diameter is 0.5 cm and the inner conductor is 0.11 cm in diameter. What is the surge impedance?

$$Z_0 = 138 \log \left(\frac{D_S}{D_C} \right) = 138 \log \left(\frac{0.5}{0.11} \right)$$

$$= 138 \log (4.545) = 138 \times 0.6576$$

$$= 91 \ \Omega$$

EXAMPLE 95-2

Compute the surge impedance of a larger type of coaxial cable with a D_S of 1 cm and a D_C of 0.22 cm. Note the comparison to the answer in Example #1.

$$Z_0 = 138 \log \left(\frac{D_S}{D_C} \right) = 138 \log \left(\frac{1}{0.22} \right)$$

$$= 138 \log 4.545 = 138 \times 0.6576$$

$$= 91 \, \Omega$$

EQUATION 96
RF WAVELENGTH

$$\lambda = \frac{300}{f}$$

λ = wavelength   m
f = frequency   MHz
300 = velocity constant   m/s

Signals whose frequency is below 10 kHz are basically confined to traveling along fixed conductors. Above this frequency the waves can emanate into free space as radio-frequency waves. This action occurs because the very rapid oscillations cannot collapse back upon themselves but are pushed outward from a radiator, such as an antenna. Since these RF waves are traveling at 3×10^8 m/s, we may determine the physical length in meters of each wave by dividing 300 by the frequency in megahertz. The answer obtained will be the exact length of a single cycle. For instance, at 15 MHz the wavelength is 20 m, and at 30 MHz, 10 m. Ham radio operators refer to radio bands at these frequencies as the 20- and 10-meter bands.

Antenna height too may be expressed in wavelengths. This is accomplished by first determining the antenna height in meters and then dividing it by the metric length of the wave being transmitted.

EXAMPLE 96-1

Calculate the wavelength of a 750-kHz signal.
First express the frequency in MHz:

$$750 \text{ kHz} = 0.75 \text{ MHz}$$

Then determine λ:

$$\lambda = \frac{300}{f} = \frac{300}{0.75} = 400 \text{ m}$$

EXAMPLE 96-2

A vertical antenna is 500 ft high and is transmitting a 1.3-MHz signal. Determine the physical height of the antenna in wavelengths.
1 m equals 3.28 ft. Thus,

$$\frac{500}{3.28} = 152.44 \text{ m}$$

Now find the signal wavelength:

$$\lambda = \frac{300}{f} = \frac{300}{1.3} = 230.8 \text{ m}$$

The ratio is 152.44 ÷ 230.8, or 0.66. In other words, the antenna is about two-thirds of a wavelength at 1.3 MHz.

EQUATION 97
ANTENNA LENGTH

$$L = \frac{468}{f}$$

L = antenna length   ft
f = signal frequency   MHz
468 = velocity constant   ft/s

In general practice, antennas are constructed to be either one-fourth or one-half the wavelength of the signal. Whether the signal is to be transmitted or received by the antenna is of no difference—the same calculations apply. Equation 97 computes the actual physical length of a half-wave radiator. The classic design is the one-half center-feed dipole shown in Fig. 97-1.

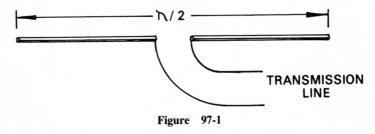

Figure 97-1

Equation 97 is derived directly from Equation 96 as follows:

$$\lambda = \frac{3 \times 10^8 \text{ m/s}}{f \text{ (Hz)}}$$

$$\frac{\lambda}{2} = \frac{984 \text{ ft/s}}{f \text{ (MHz)}} \div 2 = \frac{492 \text{ ft/s} \times 0.95}{f \text{ (MHz)}}$$

$$= \frac{468 \text{ ft/s}}{f \text{ (MHz)}}$$

You will notice there is a correction factor of 0.95 in the conversion. This accounts for the antenna dielectric's signal retardation. A signal cannot travel as fast along the antenna as it can in free air; therefore, we need the correction factor.

EXAMPLE 97-1

We wish to make a half-wavelength folded dipole antenna from 300-Ω twinlead for our FM receiver. Determine the correct antenna length for this purpose. (See Fig. 97-2.)

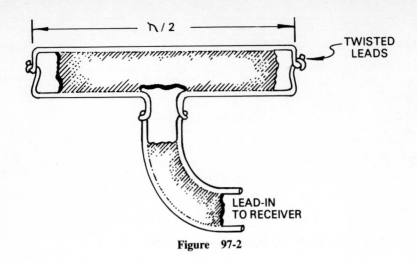

TWISTED
LEADS

LEAD-IN
TO RECEIVER

Figure 97-2

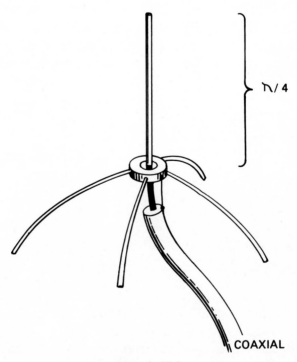

Figure 97-3

Since the center of the FM band is approximately 100 MHz, we will use this value.

$$L = \frac{468}{f} = \frac{468}{100} = 4.68 \, \text{ft}$$

EXAMPLE 97-2

Determine the physical length of a ¼-wavelength radiator for the 30-MHz ground plane antenna in Fig. 97-3.

$$L = \frac{468}{f} = \frac{468}{30}$$

= 15.6 ft (for a half-wavelength radiator)

½ L = 7.8 ft (for a quarter-wavelength radiator)

EQUATION 98
DIRECTIONAL ANTENNA DESIGN

$X = 0.25\lambda$	X = reflector spacing	in, cm*
$Y = 0.2\lambda$	Y = director spacing	in, cm*
	λ = signal wavelength	in, cm*

There are numerous types of directional antenna designs that will provide an increase in gain over what a dipole (the standard) will provide. The one we shall cover here is the parasitic Yagi type. This antenna consists of a half-wavelength driven element (DE) preceded by a director rod and followed by a reflector rod as shown in Fig. 98-1. The total length of the driven element is $\frac{1}{2}\lambda$ and is found by using Equation 97. The director is 5 percent shorter, while the reflector is 5 percent longer. Spacing between elements is critical and can be approximated with $X = 0.25\lambda$ and $Y = 0.2\lambda$. After construction is complete, it is suggested that you prune the element lengths and make slight spacing adjustments as follows: Connect the antenna to a low-power transmitter and at a distance of at least 10 wavelengths monitor the signal on the S meter of a good receiver. Such an antenna should have an 8.2-dB gain over that of a dipole.

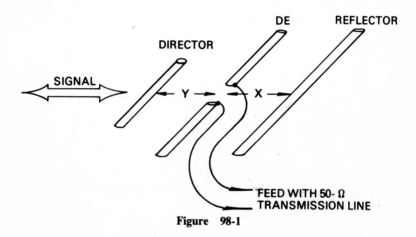

Figure   98-1

EXAMPLE 98-1

In designing a Yagi for 27 MHz, how long should the driven element be?

*Units must be similar.

Also calculate the lengths of the director and the reflector.

$$L = \frac{468}{f} = \frac{468}{27}$$

= 17.33 for the driven element (split and driven at its center)

17.33 × 0.95 = 16.46 ft for the director
17.33 × 1.05 = 18.20 ft for the reflector

EXAMPLE 98-2

For the antenna in Example 98-1, what should the element spacing be? If $\lambda/2 = 17.33$ ft, then $\lambda = 34.66$ ft.

$X = 0.25\lambda = 0.25 \times 34.66$

= 8.7 ft from reflector to driven element

$Y = 0.2\lambda = 0.2 \times 34.66$
= 6.9 ft from director to driven element

NOTE It can be seen that it would not be physically practical to design such an antenna for lower frequencies. Also, if the antenna is to be used for citizens band operation it should be vertically polarized (placed on its side with the elements running vertically).

EQUATION 99
LOCAL OSCILLATOR FREQUENCY

$$f_{LO} = f_s \pm f_{IF}$$

f_{LO} = oscillator frequency   Hz*

f_s = signal frequency   Hz*

f_{IF} = intermediate frequency   Hz*

When two signals at different frequencies are combined, a total of four basic frequencies are produced. The first two are the original ones that "beat" together; the other two are the sum and difference frequencies. As an example, beating 1 MHz against 1.455 MHz yields 1 MHz, 1.455 MHz, 2.455 MHz, and 0.455 MHz. This action is often referred to as the *superheterodyne principle* when applied to a radio receiver.

Within a receiver is a local oscillator that is always tuned 455 kHz above the desired signal. Therefore, one of the four signals resulting from beating will always be 455 kHz itself (the difference frequency). This particular frequency is filtered from the other three by tuned-circuit filtering and is passed on to a 455-kHz tuned amplifier. Very efficient tuned circuits can be used in the receiver, since they will always be dealing with the same 455-kHz frequency regardless of the desired signal.

In all standard AM receivers the local oscillator is at a frequency above that of the incoming signal. In other types of communications equipment this frequency may be below the incoming signal. In either case, the difference frequency will always remain the same.

EXAMPLE 99-1

What should the frequency of a local oscillator be to process a 1.2-MHz signal down to an intermediate frequency of 455 kHz?

First convert 455 kHz to MHz:

$$455\text{ kHz} = 0.455\text{ MHz}$$

Then compute f_{LO}:

$$f_{LO} = f_s + f_{IF} = 1.2\text{ MHz} + 0.455\text{ MHz} = 1.655\text{ MHz}$$

*Similar units such as kHz or MHz must be used.

EXAMPLE 99-2

If the standard AM broadcast band is from 550 kHz to 1.65 MHz, what must the range of the local oscillator be?

Low end		High end
$f_{LO} = f_s + f_{IF}$	f_{LO}	$= f_s + f_{IF}$
$= 0.550\,\text{MHz} + 0.455\,\text{MHz}$		$= 1.65\,\text{MHz} + 0.455\,\text{MHz}$
$= 1.005\,\text{MHz}$		$= 2.105\,\text{MHz}$

Thus, the local oscillator frequency must have a range of 1.005 MHz to 2.105 MHz in order to maintain the 455-kHz IF.

EQUATION 100
IMAGE FREQUENCY

$$f_{IMG} = f_s + 2f_{IF}$$

f_{IMG} = image frequency   Hz*
f_s = signal frequency   Hz*
f_{IF} = intermediate frequency   Hz*

In the standard "superhet" receiver the local oscillator is set 455 kHz above the incoming signal. Beating these together produces an intermediate frequency at 455 kHz. For instance, if the station's signal is at 600 kHz, the local oscillator will be at 1055 kHz, so that the IF is 455 kHz. The 600-kHz station, however, has an image frequency of 1510 kHz. If another station of sufficient strength is operating at this frequency, the receiver may process the higher signal as well. It can be seen that 1510 kHz minus the local oscillator frequency of 1055 kHz will also produce the 455-kHz IF. To overcome this problem, the front end of the receiver must have a sufficiently high-Q tuned circuit (recall Equation 56). It will be tuned to the lower (600-kHz) station and not the higher one. Upper-frequency rejection will now result. The image frequency, then, is the main signal frequency plus twice the value of intermediate frequency, as indicated by Equation 100.

EXAMPLE 100-1

What is the image frequency of a 730-kHz signal if the intermediate frequency is 455 kHz?

$$f_{IMG} = f_s + 2f_{IF} = 730 \text{ kHz} + 2(455 \text{ kHz})$$
$$= 730 \text{ kHz} + 910 \text{ kHz} = 1640 \text{ kHz}$$

EXAMPLE 100-2

If the range of the standard AM band is 550 to 1650 kHz, will the image frequency of a 1.2-MHz signal be of any concern?

$$f_{IMG} = f_s + 2f_{IF} = 1200 \text{ kHz} + 2(455 \text{ kHz})$$
$$= 1200 \text{ kHz} + 900 \text{ kHz} = 2100 \text{ kHz}$$

Since this upper frequency is well beyond the upper band limit, it should be of no consequence.

*Similar units such as kHz or MHz must be used.

BIBLIOGRAPHY

Angerbauer, G. J.: *Electronics for Modern Communications,* Prentice-Hall, Inc., Englewood Cliffs, N.J.

Bannon, E.: *Operational Amplifiers,* Reston Publishing Company, Reston, Va., 1975.

Coughlin, R. F., and F. F. Driscol: *Operational Amplifiers and Linear Integrated Circuits,* Prentice-Hall, Inc., Englewood Cliffs, N.J., 1977.

Grob, B.: *Basic Electronics,* McGraw-Hill Book Company, New York, 1977.

Herrington, D., and S. Meacham: *Electronic Tables and Formulas,* Howard W. Sams & Co., Inc., Indianapolis, 1973.

Metzger, D. D.: *Electronic Circuit Behavior,* Prentice-Hall, Inc., Englewood Cliffs, N.J., 1975.

Pettit, M. P., and M. M. McWhorter: *Electronic Switching, Timing and Pulse Circuits,* McGraw-Hill Book Company, New York, 1970.

Prensky, S. D.: *Electronic Instrumentation,* Prentice-Hall, Inc., Englewood Cliffs, N.J., 1971.

Schrader, R. L.: *Electronic Communication,* McGraw-Hill Book Company, New York, 1975.

Signetics Digital, Linear, MOS Applications Manual, Signetics Corp. 1974

Tocci, R. J.: *Fundamentals of Electronic Devices,* Charles E. Merrill Publishing Company, Columbus, Ohio, 1975.

INDEX TO TERMS

INDEX TO EQUATION TITLE

INDEX TO EQUATIONS

$$a = \sqrt{\frac{Z_P}{Z_L}}$$ **112**

$$A_V = g_{fs} \times R_L$$ **134**

$$A_V = \frac{\mu \times R_L}{r_p + R_L}$$ **137**

$$A_V = \frac{R_C \| R_L}{R_E}$$ **156**

$$A_V = \frac{R_i + R_F}{R_I}$$ **159**

$$B = \frac{\phi}{A}$$ **35**

$$BW = \frac{f_0}{Q}$$ **110**

$$C = \frac{1}{2\pi f X_C}$$ **90**

$$C = \frac{I \times T}{V}$$ **150**

$$C = \frac{1 \times 10^6}{2\pi f(R \times 0.1)}$$ **176**

$$C_T = \frac{1}{1/C_1 + 1/C_2 + \ldots + 1/C_n}$$ **52**

$$C_T = C_1 + C_2$$ **54**

$$f = \frac{1}{T}$$ **72**

$$f = \frac{P \times S}{120}$$ **75**

$$f = \frac{1}{RC \times \ln/(V_{IN} - V_{P3})}$$ **130**

$$f = \frac{1.44}{(R_A + 2R_B)C}$$ **162**

$$f = \frac{2(V_{CC} - V_C)}{R \times C \times V_{CC}}$$ **164**

$$f_0 = \frac{1}{2\pi \sqrt{LC}}$$ **108**

$$f_s = -k \times \Delta t \times f$$ **184**

$$f_{CO} = \frac{R}{2\pi L}$$ **104**

$$f_{CO} = \frac{1}{2\pi RC}$$ **106**

$$f_{IMG} = f_S + 2f_{IF}$$ **202**

$$f_{LO} = f_s \pm f_{IF}$$ **200**

$$g_m = \frac{\Delta I_p}{\Delta V_g}$$ **136**

$$g_{fs} = \frac{\Delta I_{DP}}{\Delta V_{GS}}$$ **132**

$$I = \frac{Q}{T}$$ **2**

$$I = \frac{V}{R} \quad \textbf{6}$$

$$I_C = \beta \times I_B \quad \textbf{124}$$

$$I_E = I_B + I_C \quad \textbf{122}$$

$$I_L = \frac{V_{TH}}{R_L + R_{TH}} \quad \textbf{28}$$

$$I_{BR} = \frac{R_{OP}}{R_1 + R_2} \times I_T \quad \textbf{26}$$

$$I_{SAT} = \frac{V_{CC}}{R_C} \quad \textbf{146}$$

$$L = \mu \times \frac{N^2 \times A}{l} \times 1.26 \times 10^{-6} \quad \textbf{38}$$

$$L \cong \frac{1400}{2\pi f} \quad \textbf{172}$$

$$L = \frac{468}{f} \quad \textbf{195}$$

$$L_m = k \sqrt{L_1 L_2} \quad \textbf{44}$$

$$L_T = L_1 + L_2 \ldots + L_n \quad \textbf{40}$$

$$L_T = \frac{1}{1/L_1 + 1/L_2 + \cdots + 1/L_n} \quad \textbf{42}$$

$$L_T = L_1 + L_2 \pm 2L_m \quad \textbf{46}$$

$$\lambda = \frac{3 \times 10^8}{f} \quad \textbf{77}$$

$$\lambda = \frac{300}{f} \quad \textbf{193}$$

$$N = \frac{\sqrt{L(9r + 10l)}}{r} \quad \textbf{174}$$

$$N_{dB} = 10 \times \log \frac{P_{OUT}}{P_{IN}} \quad \textbf{170}$$

$$\frac{N_P}{N_S} = \frac{V_P}{V_S} \quad \textbf{81}$$

$$P = V \times I \quad \textbf{17}$$

$$P = I^2 R \quad \textbf{19}$$

$$P = \frac{V^2}{R} \quad \textbf{21}$$

$$P_A = V \times I_T \quad \textbf{100}$$

$$PF = \frac{P_T}{P_A} \quad \textbf{102}$$

$$P_T = I^2 R \quad \textbf{98}$$

$$P_{SB} = \frac{M^2}{2} \times P_C \quad \textbf{186}$$

$$\%M = \frac{\Delta f}{75 \times 10^3 \text{ Hz}} \times 100 \quad \textbf{187}$$

$$Q = I \times T \quad \textbf{4}$$

$$Q = C \times V \quad \textbf{48}$$

$$Q = \frac{X_L}{R} \quad \textbf{87}$$

$$R = \rho \frac{l}{A} \quad \textbf{31}$$

$$R_B = \frac{V_{IN} - V_{BE}}{2 \times I_B} \quad \textbf{152}$$

$$RF = \frac{\text{rms ripple}}{V_{DC}} \quad \textbf{148}$$

$$R_i = \frac{V_{NL} - V_L}{I_L} \quad \textbf{33}$$

$$R_1 = \frac{2}{2\pi \times \Delta f \times C} \quad \textbf{178}$$

$$R_S \cong V \times S \quad \textbf{63}$$

$$R_T = R_1 + R_2 \ldots + R_n \quad \textbf{10}$$

$$R_T = \frac{R_1 \times R_2}{R_1 + R_2} \quad \textbf{12}$$

$$R_T = \frac{1}{1/R_1 + 1/R_2 + \ldots + 1/R_n} \quad \textbf{15}$$

$$R_{SH} = \frac{R_m \times I_m}{I_T - I_m} \quad \textbf{60}$$

$$R_{ZH} = \frac{V_S - V_Z}{I_L + I_{ZK}} \quad \textbf{118}$$

$$R_{ZL} = \frac{V_S - V_Z}{0.5 \times I_{ZM}} \quad \textbf{120}$$

$$S = \frac{1}{I_m} \quad \textbf{62}$$

$$S = \frac{f \times 120}{P} \quad \textbf{76}$$

$$T = \frac{L}{R} \quad \textbf{56}$$

$$T = RC \quad \textbf{58}$$

$$T = \frac{1}{f} \quad \textbf{74}$$

$$T = 1.1 \times RC \quad \textbf{161}$$

$$T = NI \quad \textbf{37}$$

$$t = RC \ln \left(1 - \frac{v}{V} \right) \quad \textbf{168}$$

$$V = I \times R \quad \textbf{8}$$

$$V_{OUT} = \frac{R_X}{R_T} \times V_S \quad \textbf{23}$$

$$V_S = V_L + V_F \quad \textbf{116}$$

$$V_{CO} = \frac{-V_B}{\mu} \quad \textbf{139}$$

$$v = V_{MAX} \times \sin \theta \quad \textbf{66}$$

$$v = V(1 - \epsilon^{-t/RC}) \quad \textbf{166}$$

$$\frac{V_P}{V_S} = \frac{I_S}{I_P} \quad \textbf{83}$$

$$V = \frac{Q}{C} \quad \textbf{50}$$

$$X_C = \frac{1}{2\pi f C} \quad \textbf{88}$$

$$V_P = \eta\, V_{BB} + V_{DD} \quad \textbf{128}$$

$$X_L = 2\pi fL$$
or \hspace{3cm} **85**
$$X_L = \omega L*$$

$$V_Y = \sqrt{3} \times V \quad \textbf{79}$$

$$Z = \sqrt{R^2 + X^2} \quad \textbf{92}$$

$$V_{CC} = I_C \times R_C + V_{CE} \quad \textbf{126}$$

$$Z = \frac{V}{I_T} \quad \textbf{95}$$

$$V_{DT} = V_R + V_F \quad \textbf{142}$$

$$V_{GS} = -V_{RS} \quad \textbf{144}$$

$$Z_o = R_C \| R_L \quad \textbf{154}$$

$$V_{AV} = 0.637 \times V_{MAX} \quad \textbf{68}$$

$$Z_0 = 276 \log\left(\frac{D}{R}\right) \quad \textbf{189}$$

$$V_{RMS} = 0.707 \times V_{MAX} \quad \textbf{70}$$

$$Z_0 = 138 \log\left(\frac{D_S}{D_C}\right) \quad \textbf{191}$$

$$V_{MAX} = 1.414 \times V_{RMS} \quad \textbf{71}$$